W0037325

Fortschritte der Chemie organischer Naturstoffe

Progress in the Chemistry of Organic Natural Products

52

Founded by L. Zechmeister
Edited by W. Herz, H. Grisebach, G.W. Kirby,
and Ch. Tamm

Authors:
H. Achenbach, P. Bhattacharyya,
D. P. Chakraborty, T. Goto, L. Merlini,
G. Nasini, U. Weiss

Springer-Verlag
Wien New York 1987

Dr. W. Herz, Professor of Chemistry, Department of Chemistry,
The Florida State University, Tallahassee, Florida, U.S.A.

Prof. Dr. H. Grisebach, Biologisches Institut II, Lehrstuhl für Biochemie der Pflanzen,
Albert-Ludwigs-Universität, Freiburg i. Br., Federal Republic of Germany

G.W. Kirby, Sc. D., Regius Professor of Chemistry, Chemistry Department,
The University, Glasgow, Scotland

Prof. Dr. Ch. Tamm, Institut für Organische Chemie der Universität Basel,
Basel, Switzerland

With 65 Figures

This work is subject to copyright.
All rights are reserved, whether the whole or part of the material is concerned, specifically those
of translation, reprinting, re-use of illustrations, broadcasting, reproduction by photocopying
machine or similar means, and storage in data banks.

© 1987 by Springer-Verlag/Wien
Softcover reprint of the hardcover 1st edition 1987
Library of Congress Catalog Card Number AC 39-1015

ISBN-13: 978-3-7091-8908-5   e-ISBN- 978-3-7091-8906-1

DOI: 10.1007/ 978-3-7091-8906-1

Contents

List of Contributors

ACHENBACH, Dr. H., Institute for Pharmacy and Food Chemistry, University of Erlangen, D-8520 Erlangen, Federal Republic of Germany.

BHATTACHARYYA, Dr. P., Department of Chemistry, Bose Institute, 93/1 Acharya Prafulla Chandra Road, Calcutta 700 009, India.

CHAKRABORTY, Dr. D.P., Department of Chemistry, Bose Institute, 93/1 Acharya Prafulla Chandra Road, Calcutta 700 009, India.

GOTO, Professor T., Laboratory of Organic Chemistry, Faculty of Agriculture, Nagoya University, Chikusa, Nagoya 464, Japan.

MERLINI, Dr. L., Istituto di Biochimica e di Chimica, Facoltà di Agraria, Università di Milano, Via Celoria, 2, I-20133 Milano, Italy.

NASINI, Dr. G., Centro del C.N.R. per le Sostanze Organiche Naturali, Dipartimento di Chimica, Politecnico di Milano, I-20133 Milano, Italy.

WEISS, Dr. U., Laboratory of Chemical Physics, National Institute of Diabetes and Digestive and Kidney Diseases, National Institutes of Health, Bethesda, MD 20892, U.S.A.

Naturally Occurring Perylenequinones

By U. WEISS, Laboratory of Chemical Physics, National Institute
of Diabetes and Digestive and Kidney Diseases, National Institutes
of Health, Bethesda, Maryland, U.S.A.

L. MERLINI, Istituto di Biochimica e di Chimica, Facoltà di Agraria,
Università di Milano, Italy

G. NASINI, Centro del C.N.R. per le Sostanze Organiche Naturali,
Dipartimento di Chimica, Politecnico di Milano, Italy

With 6 Figures

Contents

1. Introduction

Perylenequinones, together with a few compounds containing a partially reduced perylene skeleton, form a relatively small but rapidly increasing group of chemically interesting, biologically active pigments obtainable from natural sources.

The present review intends to bring together the available information on these substances in comprehensive fashion; no review of the entire topic has been available so far. The literature has been collected, as completely as possible, through 1985. The review consists of a general introduction (Section 2.), monographic sections on individual compounds (Sections 3. and 8.), and general discussions on such more important topics as stereochemistry (Section 4.), tautomerism (Section 5.), biosynthesis (Section 6.), and biological activity (Section 7.).

2. Natural Perylenequinones in General

2.1. Perylene and Its Derivatives

The parent, fully aromatic hydrocarbon perylene (1) was first obtained synthetically by SCHOLL et al. in 1910 (118), through heating of 1,1'-dinaphthyl, or of naphthalene itself, with AlCl$_3$, or by Ullmann reaction of 1,8-diiodonaphthalene. It was subsequently found to be a minor constituent of coal tar (49). The chemistry of the perylene system was later explored by ZINKE, who described his research in a long series of papers in *Monatsh. Chem.* from 1919 on; compounds obtained by him include several quinones, particularly the yellow 3,10-quinone (2) (146), from which all the naturally occurring perylenequi-

(1) (2) (3)

nones are derived (for a possible exception, see the discussion of the pigments from the mineral quincyte in Section 2.2.). Its 4,9-dihydroxy derivative (3) was likewise first prepared by ZINKE (*145*), but its structure was recognized only later (*30*). It is the simplest of the naturally occurring perylenequinones (see Section 3.) and the parent compound of all the others, with the possible exception of some of the mold pigments discussed in Section 3.7. and of the pigments of quincyte (see below).

2.2. Types of Perylenequinones Obtainable from Natural Sources. Occurrence

Besides fully aromatic perylenequinones, a few natural compounds with partially reduced perylene skeleton have been encountered; because of their close chemical and presumably biosynthetic relationship with the true perylenequinones they are included in this review (see Section 8.).

Most of the natural perylenequinones fall into one of three classes:

(A) C_{20} Compounds Without Carbon Substituents. This class includes the parent compound (3), which occurs in certain molds (see Section 3.1.), and some partially reduced compounds produced by molds of the genera *Alternaria* and *Stemphylium* (see Section 8.).

(B) Mold Perylenequinones Carrying Carbon Substituents. Almost all of the pigments in this class (the majority of perylenes from molds) are derivable from the unit of diagram 1; phleichrome (Section 3.5.) and cercosporin (Section 3.4.) are the most important ones. In some pigments of this class (the elsinochromes, Section 3.3., and hypocrellin, Section 3.8.) additional alicyclic rings are present, no doubt formed through interaction of the two C_3 side-chains at some stage of the biosynthesis. In two structures otherwise belonging into this class the

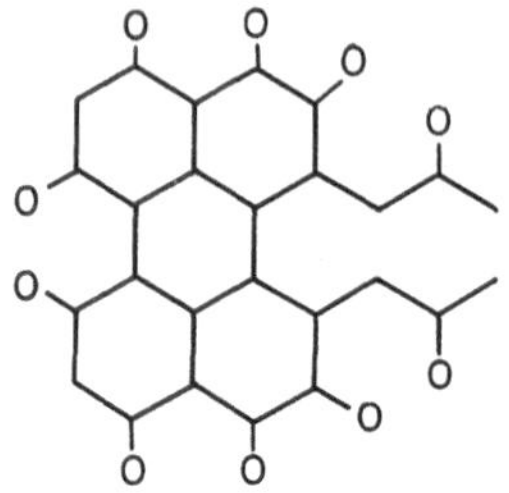

Diagram 1

positions of one or two of the oxygenated substituents on the chromophore may be different from the ones shown in diagram 1 (see Section 3.7.).

The pigments of this class are produced by a wide variety of molds, most of them phytopathogens. It seems that the damage done to the host plants by these molds is caused by the photodynamic action of their perylenequinones (see Section 7.).

It will be noted that the substituents on the perylene system of diagram 1 are arranged in a bilaterally symmetrical fashion (symmetry group C_{2v}), which contrasts from the arrangement found in the next class.

Almost all the pigments of class B are derivatives of (3) in which both phenolic hydroxyls are free, so that both quinone carbonyls are chelated; elsinochrome D, a complex monoether with only *one* free hydroxyl and possibly two pigments of Section 3.7. are exceptions. The presence of the methoxyls in positions 5 and 8 of the perylene system, found in all compounds of this group, is biosynthetically unusual.

(C) Perylenequinones from Aphids: Rhodoaphin-be; Erythroaphins-fb and -sl. The pigments of this group can be isolated from aphids although only rhodoaphin (as the incompletely investigated glycoside heteroaphin) preexists in the insects, while the two erythroaphins are actually formed by post-mortem stepwise enzymatic transformations from native precursors, the protoaphins, which do not yet contain the ring system of perylene (see Section 3.9.). Neither the two stereoisomeric erythroaphins nor certain intermediates in their formation, the xantho- and chrysoaphins, can thus qualify as *bona fide* natural compounds; they might therefore not have been included in this review of *natural* perylenequinones. They were, however, the first derivatives of perylene to be isolated from natural material, and the seminal research by Lord Todd and coworkers, which led from 1948 on to the establishment of their structures and the clarification of their reactions, has provided most of the methodology used in subsequent work by other investigators *(43)*.

Erythroaphin-*fb*	(**4**) R = R′ = αH
Erythroaphin-*sl*	(**5**) R = βH; R′ = αH
Rhodoaphin-*be*	(**6**) R = R′ = βOH

The general structure of the three pigments, erythroaphin-*fb* (**4**) erythroaphin-*sl* (**5**), and rhodoaphin-*be* (**6**) exhibits a centrosymmetric arrangement (group C_{2h}) of the substituents around the perylenequinone nucleus.

The color of quincyte, a pink variety of the silicate mineral sepiolite or meerschaum, seems to be caused by derivatives of perylene (apparently, however, not of 4,9-dihydroxy-perylene-3,10-quinone), which were examined by WATTS *et al.* (*134*). For isolation of these pigments treatment of quincyte with 20% aqueous HF is required. Chromatography gave four orange fractions having essentially identical absorption spectra, which remain unchanged by addition of aqueous NaOH. The least polar fraction, apparently $C_{26}H_{20}O_4$ (mass spectrometry), gave a diacetate with a spectrum very similar to the one of perylene; this diacetate was also obtained on attempted reductive acetylation or boroacetylation. Hydrolysis and subsequent oxidation of the diacetate gave an orange compound interpreted as a quinone, with an absorption spectrum between ca. 400 and 500 nm which resembles the group of three bands between 500 and 600 nm in the spectrum of erythroaphin-*fb*.

In a later study (*131*) one pigment from quincyte was obtained as reddish-brown sublimable crystals of molecular weight 396, IR (KBr) 1645 cm^{-1}. The absorption spectrum of this pigment agrees well with published (*134*) data. On the basis of the very simple ^{1}H NMR spectrum, the unusual structure (**7**) or less probably (**8**) was proposed

(**7**)   (**8**)

for this pigment. If this constitution is correct, the formation of the diacetate described by Watts *et al.* (*134*) must involve a redox reaction. However, the reported (*134*) lack of quinonoid properties of the main pigment would be difficult to reconcile with structure (**7**), provided the crystalline pigment of Treibs *et al.* (*131*) is identical with the substance of Watts. No direct comparison of the samples is mentioned.

In conclusion, it may be stated that the pigments of quincyte may be indeed derived from perylene, but that this relationship has not been firmly established, and that structure (**7**) should be considered provisional. Reinvestigation of the problem seems highly desirable.

It seems surprising that perylenequinones should occur in molds and in one class of insects, but be almost completely lacking in higher plants (unless our knowledge should still be too incomplete to give a trustworthy picture). We have encountered only one single observation which suggests presence of such a pigment in a higher plant: *Diospyros natalensis* susp. *natalensis* (Ebenaceae) contains (*132*) a pigment obtained as brown-red needles, which shows spectroscopic properties strongly suggesting a derivative of (**3**): electronic spectrum similar to the one of (**3**), $v(CO)$ 1623 cm^{-1}. Unfortunately, this interesting substance has not been completely characterized: no m.p., elementary analysis or mass-spectrometric molecular weight. On the basis of the ^{1}H NMR spectrum consisting of five singlet peaks in approximate intensity ratios 1:1:3:3:3, and of their chemical shifts, the unusual structure of 4,9-dihydroxy-1,2,6,7,11,12-hexamethylperylene-3,10-quinone has been proposed for this unique substance.

2.3. General Properties and Reactions of Natural Perylenequinones

The properties and reactions discussed in this part of the review are documented by a few selected examples. No complete coverage is attempted.

The fully chelated 4,9-dihydroxyperylene-3,10-quinones, which constitute the large majority of compounds pertinent to the review, are bright red pigments. Their red solutions in organic solvents show intense red fluorescence. The fluorescence spectra of some of them (elsinochrome A, cercosporin, isocercosporin, phleichrome) have been investigated (*139*). The solubility of some of these pigments (e.g. (**3**)) in the common organic solvents is very low while others are quite soluble. In basic media (NaOH, Na$_2$CO$_3$, NH$_3$, aqueous pyridine) these pigments dissolve with a bright green color, a behavior characteristic of chelated extended quinones, i.e. those in which the quinonoid grouping extends over more than one ring; other phenolic red pigments tend to give purple to blue solutions in alkali and the like. Derivatives of

(3) in which only one carbonyl is chelated (e.g. elsinochrome D and the monomethyl ethers of elsinochrome A, described in Section 3.3.) are orange and give brownish-green solutions in alkali. Non-phenolic substances of this type, such as (2), are yellow (cf. (61)); the formation of red dimethyl ethers of elsinochrome A (81), phleichrome and iso-phleichrome (4) (λ_{max} 564, 570 nm, resp.) is a surprising exception, un-explained to the best of our knowledge. The isomeric dimethyl ethers derived from the other tautomer of the parent compounds (see Section 5.) are yellow, as expected.

2.4. Spectroscopic Properties

The many-banded electronic spectrum in the visible region of (3) and its fully chelated derivatives is highly characteristic and diagnostic; as an example, the spectra of elsinochrome A and of erythroaphin-*fb* are shown in Fig. 1 of Section 3.3.

The infrared spectra of the perylenequinones show carbonyl stretching bands at the low frequencies characteristic for extended quinones. Non-chelated quinones of the perylene series, e.g. (2), show $v(CO)$ at 1650 cm^{-1} (25); in both isomers of the dimethyl ether of elsino-chrome A this band occurs at 1634 cm^{-1} (80), in the "normal" di-methyl ether of (4) at 1638 cm^{-1} (34). The red and yellow dimethyl ethers of isophleichrome, however, show it at the surprisingly low fre-quency of 1610 cm^{-1} (4). Chelation moves the band to somewhat lower frequencies; cf. $v(CO)$ 1631 cm^{-1} for (3) (30), 1623 cm^{-1} for both elsinochrome A (80) and cercosporin (140).

The absence or weakness of an OH stretching frequency from the hydroxyls on the chromophore in elsinochrome A (80), cercosporin (79) and other related pigments is consistent with the strong chelation.

The enormous importance of NMR in the structural elucidation of natural substances needs no emphasis. ^{1}H NMR was, e.g., vital in the elucidation of the structure of the erythroaphins (39) and elsin-ochromes (80); it was almost the only method used for the study of the other compounds discovered more recently. The biosynthetic path-way leading to cercosporin (109) and elsinochromes (47, 76) was ex-plored with the help of ^{13}C NMR. The nuclear Overhauser effect has been used, e.g., in the structure proof of elsinochrome A (84), and in the study of structure and configuration of the reduced perylenequi-nones (6, 127) and of the conformation of the side chains of cercosporin (5). Some of the recently developed NMR techniques, such as two-dimensional and ^{13}C polarization transfer NMR, have found use in the structure proof of the reduced perylenequinones (6, 127).

The compact ring system of perylenequinones is rather resistant to mass-spectrometric fragmentation (17), which has therefore been used for the study of problems such as the structure and stereochemistry of the side-chains of the aphid pigments. The most important utilization of mass spectrometry, however, was the proof by high-resolution measurements that cercosporin has 29 carbon atoms (85, 140) rather than 30, as had been assumed initially (78) prior to the development of techniques for safe distinction between these two possibilities.

Examination of the ORD and CD spectra of the pigments has been important particularly in the study of the configuration of several mold pigments, and of the reduced derivatives discussed in Section 8.

Finally, X-ray crystallography has had its use in the more recent research on the natural perylenequinones. Its most productive application was probably the investigation of the well-crystallized natural monoacetate monobenzoate of cercosporin (108), which established the absolute helicity of the non-planar chromophore (see Fig. 3 in Section 4.).

2.5. Chemical Reactions

Of the many chemical reactions that have played a role in the study of constitution and stereochemistry of the natural perylenequinones, some of the more generally useful ones deserve brief discussion.

For the proof of the quinonoid nature of these pigments, their reversible reduction by sodium dithionite to leucocompounds, which appear pink to red in alkaline medium, has found the expected general use. Reductive acetylation and methylation produce yellow, strongly fluorescent and rather sensitive leuco derivatives; the absorption spectra of the leuco-tetraacetates show the expected resemblance to the one of perylene itself (see *inter alia* (24)).

The presence of the doubly chelated extended quinone system of (3) and its derivatives is demonstrated by the green color of alkaline solutions, already mentioned, and by reactions such as complexation with metals (140) or boroacetate (80). The formation of mellitic acid on energetic oxidation of erythroaphin-*fb* (70), elsinochrome A (80), aspergillin (112), and cercosporin (77, 140) with nitric acid shows the presence of the central, fully substituted benzene ring. That these compounds are derivatives of perylene is further proved by the results of distillation from zinc dust or the milder method of fusion with zinc chloride which produces perylene itself, e.g. from erythroaphin-*sl* (24) and cercosporin (140). In the compounds of group B, however, the interaction of the side chains in positions 1 and 12 usually leads to the preponderant (*140*) or exclusive (*66*) formation of 1,12-benzoperylene; in erythroaphin-*sl*, presence of carbon substituents on both sides of the molecule causes the additional appearance of coronene (*24*).

The formation of simple derivatives, such as acetates and methyl ethers, has been invaluable, in connection with NMR and mass spectroscopy, in the localization of the various OH and OMe groups.

A few substitution reactions on the perylenequinone ring have been performed, such as bromination (*23, 38*) and amination (*23*), particularly in the aphin series. Retroaldol cleavage of the 2-hydroxypropyl side chains of some mold pigments of group B has been obtained by treatment with strong alkali (*108*), whereas cyclization by sulfuric acid of the same side chains onto the adjacent methoxylated carbons has given new dihydrofuran rings.

Cycloaddition of oxygen onto the conjugated dienic system of the central ring of hypocrellin (*135*) and the phleichromes (*4*) occurs readily with formation of more or less stable endoperoxides.

2.6. Synthesis

Almost no effort has been devoted to the synthesis of the natural perylenequinones. The only compound so far prepared is again the parent compound (**3**) obtained, although in low yield, by oxidation of perylene (*30, 61*). The single attempt toward the synthesis of functionalized derivatives is that of DALLACKER and LEIDIG (*50*) who prepared the tetramethoxyquinone (**10**) from the aldehyde (**9**) in several steps.

A diglycoside of erythroaphin-*fb* (**4**) (actually derived from the 3,9-tautomer of the perylenequinone: see Section 5.) has been prepared by oxidative coupling (neutral ferricyanide) of two molecules of glycoside B, one of the two products of reductive cleavage of the protoaphins (see Section 3.9.); acidic hydrolysis gave erythroaphin itself (*32*). These

(9)   (10)

reactions constitute a partial, but not a total synthesis of erythroaphin, since glycoside B has not been synthesized so far.

3. Individual Natural Perylenequinones

3.1. 4,9-Dihydroxyperylene-3,10-quinone

4,9-Dihydroxyperylene-3,10-quinone (3), the parent of most other natural perylenequinones, was first prepared synthetically in 1929 by ZINKE (145) who, however, formulated the compound as the 3,4,9,10-diquinone. The correct structure was established later by CALDERBANK, JOHNSON, and TODD (30) in a detailed description of the chemical and spectroscopic properties of the compound. Its natural occurrence was noted for the first time by ANDERSON and MURRAY (3); it was isolated by these authors from the large black fruiting bodies of the ascomycete *Daldinia concentrica* by extraction with acetone and sublimation of the crude pigment in high vacuum. The quinone has also been obtained (59) from the smaller fruiting bodies of another fungus, *Bulgaria inquinans*, where it occurs together with two other quinonoid pigments, bulgarein (12) and bulgarhodin (13) which are derivatives of benzo(*i*)fluoranthene.

Detailed study of the pigments of *Daldinia concentrica* by ALLPORT and BU'LOCK (1, 2) has shown that dihydroxyperylenequinone occurs in this species together with 4,4',5,5'-tetrahydroxy-1,1'-dinaphthyl (11) and with a black, insoluble polymer. Since the dinaphthyl readily yields the two other products on oxidation, either chemical or enzymatic, it can hardly be doubted that it is the parent of the quinone and the polymer also *in vivo* by phenol oxidation.

(11) (12) R = H
      (13) R = OH

3.2. Aspergillin

This pigment from the black spores of the mold *Aspergillus niger* is very probably related to 4,9-dihydroxyperylene-3,10-quinone although details of its structure are still unknown. Aspergillin has been studied most recently by LUND, ROBERTSON, and WHALLEY (*86*) and by BARBETTA, CASNATI, and RICCA (*13*); those papers contain references to older work.

Aspergillin can be extracted from the spores by dilute aqueous sodium hydroxide (*86*) or ammonia (*13*). Acidification of the extract with HCl yields the crude pigment as a black, amorphous mass. Purification was achieved by dialysis (*86*) which frees the material from iron (believed initially to be an essential constituent) or by gel chromatography on Sephadex (*13*). The major still amorphous fraction obtained by the second method seems to consist of a protein carrying a black prosthetic group. Hydrolysis with 20% aqueous HCl at 100° (*13*) yielded a number of amino acids and an amorphous black fraction which retains the spectroscopic (UV, IR) properties of the starting material and gives the same degradation products. A similar product obtained (*86*) by more energetic treatment with acid (twice-repeated refluxing with 20% sulfuric acid) was free of nitrogen.

Oxidation of aspergillin with H_2O_2 or nitric acid yields mellitic acid, a result first obtained by QUILICO (*112*). The quinonoid nature of the pigment is indicated by its reduction with alkaline sodium hydrosulfite which produces a yellow leuco compound. On treatment with Raney nickel in hot (100–200°) alkaline solution, CASNATI et al. (*13*) obtained perylene together with its tetradecahydro and perhydro derivatives. The interpretation of aspergillin as a protein conjugate of an extended quinone of the perylene series is also supported by an IR band at 1613 cm^{-1}, in the frequency range typical for chelated extended quinones (cf. elsinochrome A, 1623 cm^{-1} (*80*); 4,9-dihydroxyperylene-3,10-quinone, (3), 1631 cm^{-1}) (*30*). However, the exact nature of the prosthetic group is not clear. It could not be 4,9-dihydroxyperylene-3,10-quinone itself, since the properties of the product resulting from acidic hydrolysis of aspergillin are quite different from those of that quinone: black, amorphous, instead of red and crystalline; very soluble in alkali with black color, while the authentic quinone forms a greenish-blue, sparingly soluble sodium salt (*30*) and other quinones derived from (3), c.g. the elsinochromes, give emerald green solutions; solution in aqueous hydrosulfite yellow rather than reddish brown (*30*). The nondescript electronic spectrum of aspergillin is likewise quite different from the characteristic one of dihydroxyperylenequinone (*30*) with its five distinctive bands. Furthermore, protein could hardly be attached to dihydroxyperylenequinone other than by one (or both) of the phenolic hydroxyls; if so, the color of the resulting product should be lighter than that of the parent quinone; cf., e.g., the orange monomethyl ethers of elsinochrome A (*81*). Still other puzzling features exist: e.g. the material obtained on hydrolysis of aspergillin with sulfuric acid (*86*) contains 3.5% C-methyl, yet the reductive degradation of aspergillin with Raney nickel gave perylene itself (*13*) rather than any homolog.

Since nothing further seems to have been published on the chemistry of this unusual pigment, the exact nature of its relationship with the authentic natural perylenequinones, strongly suggested by the formation of perylene on reductive degradation, remains problematical.

Formation of aspergillin in the spores of *A. niger* is inhibited by copper chelators, especially 2,4-dithiopyrimidine (*115*), and by dimethylsulfoxide (*45*). It is claimed (*114*) that aspergillin, or a very similar pigment, has marked, heat-stable proteolytic activity and is an inhibitor of many enzymes.

3.3. The Elsinochromes

3.3.1. Occurrence and Formation

The bright red pigments, elsinochrome A (**14**), B (**15**)[1] and C (**16**)[1], and the orange elsinochrome D (**17**) are produced in laboratory cultures of many, but by no means all, species of the large ascomycetous genus *Elsinoë* and the associated conidial (asexual) stage *Sphaceloma*. These phytopathogenic molds constitute the separate family Elsinoaceae of the order Myriangiales.

The elsinochrome pigments were first examined by Weiss *et al.* (*137*) in work initiated at the New York Botanical Garden and continued subsequently at the National Institutes of Health. Independently, Salemink and coworkers at the University of Utrecht investigated the pigmentation of a mold furnished to them under the designation *Phyllosticta caryae* Peck, and isolated from it a crystalline red pigment which they named phycaron (*65*, *66*). However, it was found (*138*) that this pigment is actually identical with elsinochrome A. It had been established by Jenkins and Bitancourt (*71*, *72*) that the fungus used by the Dutch workers had been misidentified in an otherwise excellent

---

[1] Actually, the elsinochromes B and C described initially were shown recently to have been mixtures of stereoisomers, B_1 and B_2, and C_1 and C_2, respectively, which are epimeric at the secondary hydroxyls of the side chains (*5*) (see below). In the sequel, the older, stereochemically inhomogeneous preparations will be designated as "B" and "C", respectively.

paper by RAND (*113*), the taxonomy of this group of molds being notoriously difficult (*73*). The supposed *P. caryae* of Rand was actually the conidial *Sphaceloma* stage of *E. randii* Jenkins and Bitancourt (*71*). The name *"phycaron"* was therefore withdrawn (*138*) and the study of the elsinochromes continued as a cooperative project of both groups.

So far, these pigments have been obtained only in shake cultures in the laboratory, where they are contained almost exclusively in the bright red mycelium. In still cultures, only small amounts of pigment are formed near the surface, although the mold grows quite abundantly under these conditions; this fact suggests the involvement of an oxygen-requiring stage in the biosynthesis of the elsinochromes. So far, the pigments have not been obtained from the lesions produced on the host plants by these molds.

The color producing species of *Elsinoë* and *Sphaceloma* which have been investigated fall mostly into two groups (*138*). In one of these elsinochrome A (**14**) is the main pigment, accompanied by small amounts of "B" and "C"; in the other group the two last-named pigments prevail and only little A is present. For production of the larger amounts of pigment needed for detailed chemical studies *E. annonae* Jenkins and Bitancourt was chosen, since it is a particularly good source of the well-crystallized, readily purified elsinochrome A; isolation in pure form of the more polar "B" and "C" proved much less easy.

Elsinochrome D (**17**), discovered only later (*82*) as a constituent of the pigment mixture of *E. annonae*[1], occurs there in very small amounts. Of several culture media tried, Czapek-Dox medium gave the (relatively) best production. Surprisingly, it has been found that it and "C" are the main pigments produced by one strain of an apparently unrelated mold, *Pyrenochaeta terrestris* (*76*), a fungus causing pink-root disease of onions. The possibility is mentioned (*76*) that this strain might constitute another instance of a misidentified mold. The co-occurrence of these two pigments is not too surprising, since they should be closely connected biosynthetically (see later); it seems strange, though, that here the usual association of "B" and "C" is absent.

3.3.2. Isolation and Purification

The perylenequinone pigments of *Elsinoë* and *Sphaceloma* spp. (usually *E. annonae* or *S. randii*) are readily extracted from the mycelium

---

[1] Small amounts of several other fluorescent products have been obtained from cultures of *E. annonae* (*83*). One of these appears to be 9,10-dimethylphenanthrene.

by acetone at room temperature. Separation of pigments A, "B", and "C" can be done by countercurrent distribution (137, 138) or by chromatography on a variety of adsorbents, sec. calcium phosphate (138) or silicic acid being most frequently used.

The separation of elsinochromes "B" and "C" into their stereoisomeric constituents, B_1 and B_2, and C_1 and C_2 resp., has been achieved (5) by repeated preparative thick-layer chromatography on silica gel plates using a mixture of ethyl acetate/benzene/formic acid (1 : 1.5 : 0.5% v/v) as eluent.

3.3.3 Structure of the Elsinochromes

The elsinochromes were recognized as derivatives of 1,2-dihydro-benzo(ghi)perylenequinone in independent and complementary work in the USA and in the Netherlands (65, 66, 80, 137). Their nature as phenolic quinones follows from their solubility in aqueous alkali, ammonia, or sodium carbonate with bright green color (137) and the reversible reduction with alkaline sodium dithionite (color change from green to pink) (137) or with zinc dust in acetic acid to a yellow, green-fluorescing leuco compound (66, 80). The green color of alkaline solutions, typical of phenolic extended quinones, directed attention to the two groups of pigments of this type that had been isolated from natural sources at that time (and, incidentally, to their strong photosensitizing action): hypericin and its congeners (21) and the erythroaphins (see Section 7.). Like the elsinochromes, these pigments have characteristic absorption spectra with several bands in the visible region. While the spectrum of the elsinochromes shows little resemblance to that of hypericin, it is remarkably similar to the one of the erythroaphins and even more so to the one of the parent 4,9-dihydroxyperylene-3,10-quinone (3) (30); see Fig. 1. Furthermore, the spectra of the leuco-acetates of elsinochrome A and erythroaphin-fb are very similar to each other (80) and show the expected resemblance to that of perylene (66) and 3,4,9,10-tetraacetoxyperylene. These findings permitted the identification of the chromophore of the elsinochromes, and thus of 20 of its 30 carbon atoms, an identification further supported by formation of complexes with boron acetate and stannic chloride (65) and oxidation to mellitic acid by nitric acid (65). A band at 1623 cm^{-1} in the IR spectrum of elsinochrome A (80) and the absence of a hydroxyl band at 3300 cm^{-1} are consistent (74) with this interpretation. A band at 1715 cm^{-1} suggested a non-conjugated carbonyl, in agreement with a positive Lieben iodoform reaction (65, 66).

The simplicity of the ^{1}H NMR spectrum, which consists of six sharp

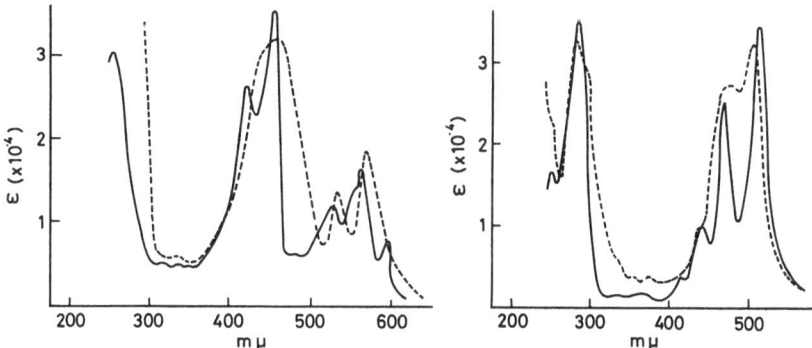

Fig. 1. Left: Absorption spectra of erythroaphin-*fb* (−) and elsinochrome A (---). Right: Absorption spectra of the leuco-acetates of erythroaphin-*fb* (−) and elsinochrome A (---). (Reproduced with permission from ref. *80*)

singlets at δ 2.05, 4.08, 4.32, 5.20, 6.63, 16.20 (in $CDCl_3$) with intensities 3:3:3:1:1:1, proved the presence of a symmetrical structure (*14, 80*). Consequently, each signal represents a pair of identical groups, identifiable by their chemical shifts as one pair of non-conjugated C-acetyls (cf. the IR band at 1715 cm^{-1}), two pairs of methoxyls, one pair of benzylic protons, one pair of aromatic or quinonoid protons, and one pair of strongly hydrogen-bonded hydroxyls. These six pairs of groups could not possibly be accommodated on the perylenequinone nucleus alone (only five binding sites available) and since two of them, the C-acetyls and the benzylic protons, are evidently not bound directly to this nucleus, a more extended carbon skeleton had to be assumed for which diagram 2 turned out to be the only expression compatible with the available evidence (*14, 66, 80*). In agreement, Clar fusion

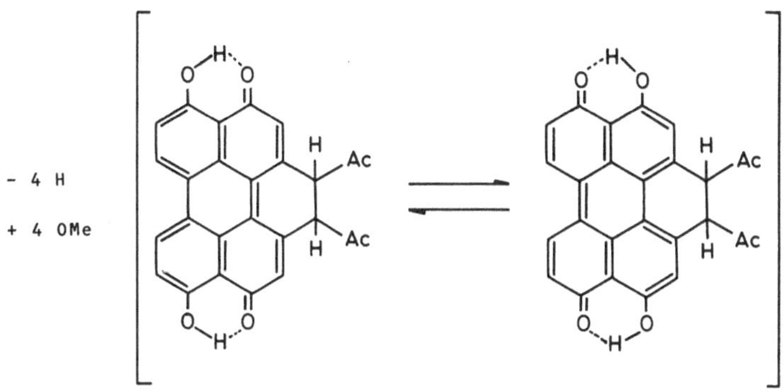

Diagram 2

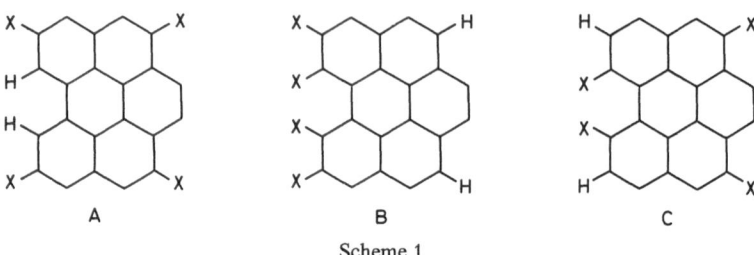

Scheme 1

with zinc dust and zinc chloride yielded (66) a product identifiable by its UV spectrum as a derivative of benzo(ghi)-perylene.

Since elsinochromes "B" and "C" are oxidized to A by chromium trioxide (66, 80), the information on the carbon skeleton of the last-named compound also applies to the two other pigments which differ from it only by reduction of one keto group, or of both, to the secondary alcohol.

There remained the question of the placement of the two pairs of methoxyls; it proved unexpectedly complex. The symmetry of the structure indicated by the NMR spectrum restricted the number of possible arrangements to the three shown in Scheme 1.

Initial attempts to interpret the absorption spectra led to structure proposals based on diagram A (14) or B (65, 66). Subsequently, however, the structure derived from C was shown (80) to be the correct one. This follows from the very ready bromination of elsinochrome A (bromine in acetic acid, 15 min, r.t.) to a dibromo derivative which lacks the NMR signals from the proton pair on the perylene system. Such easy bromination would be sterically impossible if the structure based on diagram A were correct, where the bulky bromines could be accomodated only through strong distortion of the ring system, while diagram B is disproved by the fact that in the NMR spectrum of the dibromo derivative the signals from the hydroxyls and one of the pairs of methoxyls, and only those, have shifted.

Detailed spectroscopic studies (80, 84) of elsinochrome A and its two dimethyl ethers (81) furnished conclusive proof for the correctness of structure (14) based on diagram C (Scheme 1). All the observed phenomena were entirely compatible with it and not at all with those based on diagrams A and B (Scheme 1). This is true for the upfield shifts of NMR signals from the methoxy groups upon addition of a few drops of benzene to the chloroform solution. Observations of the nuclear Overhauser effects gave the same result, as did ESR studies on the radical anions formed on electrolytic reduction of elsinochrome A and its dimethyl ethers.

a   b

(14) Elsinochrome A: $R^1 = R^2 = \text{-CO-Me}$
(15) Elsinochrome B: $R^1 = \text{-CO-Me}$, $R^2 = \text{-CHOH-Me}$
(16) Elsinochrome C: $R^1 = R^2 = \text{-CHOH-Me}$

Elsinochrome A is thus correctly represented by the two structures (14a) and (14b) in rapid tautomeric equilibrium; elsinochromes "B" and "C" are similarly formulated as (15a, b) and (16a, b) respectively.

For a discussion of the tautomerism a $\rightleftarrows$ b, generally encountered in naphthazarin (5,8-dihydroxy-1,4-naphthoquinone) and its analogs, including 4,9-dihydroxyperylene-3,10-quinone and derivatives, see Section 5.

In the case of the elsinochromes A, "B" and "C", the existence of this tautomeric equilibrium, rapid on the NMR time-scale, is demonstrated by the formation of pairs of isomeric mono- and dimethyl ethers derived from (14a) and (14b), respectively.

The two dimethyl ethers are formed, in unequal amounts, on methylation of elsinochrome A with methyl iodide and silver oxide in acetone for 15–18 hours at room temperature: a yellow isomer derived from (14b), yield ca. 50%, and a red isomer derived from (14a), yield ca. 15%. Their structures were assigned on the basis of the chemical shifts of the protons on the chromophore: yellow isomer, δ 6.13, quinonoid; red isomer, δ 6.88, benzenoid, respectively. The color of the red isomer and the surprising difference in the polarity of the two ethers remain unexplained: yellow isomer, Rf 0.59; red isomer, Rf 0.04 (81).

Both dimethyl ethers are unquestionably derived from the two tautomers a and b of elsinochrome A. A conceivable third structure derived from a 3,9-quinonoid tautomer of dihydroxyperylenequinone (see Section 5.) is excluded by all available spectroscopic evidence: a compound of this type would show signals from one benzenoid and one quinonoid proton and would have the intense absorption bands around 600 nm which are characteristic of such perylene 3,9-quinones; cf., e.g., the spectrum of the dark purple dimethyl ether of erythroaphin

(17)

fb shown in Fig. 1 of ref. (*34*). No such band occurs in the spectrum of either of the dimethyl ethers.

Interruption of the methylation reaction after seven hours gave (*81*) two orange compounds which were identified as the monomethyl ethers derived from tautomers a and b, respectively. Both these mono-ethers gave brownish-green solutions in alkali, very different from the bright green color of alkaline solutions of the elsinochromes. The IR spectrum of the isomer obtained pure showed a broad carbonyl band originating from partial overlapping of the bands of chelated (ca. 1620 cm^{-1}) and non-chelated (ca. 1630 cm^{-1}) carbonyl.

Very few such monoethers of quinones of the naphthazarin type are known; however, the orange elsinochrome D belongs to this class (*82*). It likewise gives a brownish-green solution in alkali and its IR spectrum shows a broad band with maxima at 1620 and 1634 cm^{-1}. Its absorption spectrum in the visible region resembles those of the red dimethyl ether of elsinochrome A (**14b**) and of the corresponding monomethyl ether. The complex NMR spectrum indicated a molecule lacking the symmetry of elsinochrome A. All additional evidence, in-cluding NMR solvent and nuclear Overhauser effects, showed the compound to have structure (**17**). Occurrence of a monoether of a quinone of this type among natural compounds is quite unusual, as is the pres-ence of a methylenedioxy group in such a quinone, although another variant of this group is of course prominent in cercosporin and its relatives.

3.4. Cercosporin

Cercosporin (**18**), $C_{29}H_{26}H_{10}$, was first isolated (*78*) from *Cerco-spora kikuchii*, the pathogenic fungus responsible for "purple speck disease" of soy beans, and thereafter from many other species of *Cer-*

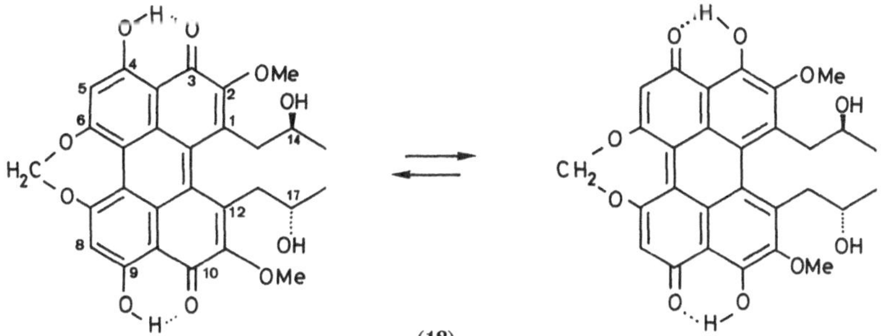

(18)

cospora (7 and refs. quoted therein). KUYAMA and TAMURA (77, 78, 79) worked out most of the structural details of cercosporin by classical redox quinone chemistry, zinc dust fusion to 1,12-benzoperylene, nitric oxidation to mellitic acid, formation of di- and tetraacetates, di- and tetramethylethers, and Thiele acetylation. From these data and the IR spectra the 4,9-dihydroxy-3,10-perylenequinone nucleus was identified; however, the required instrumental techniques (NMR, mass spectrometry) being still unavailable at the time, the correct empirical formula (C_{29} rather than C_{30}) and the complete structure could not yet be established. Later investigations (85, 140) clarified these details mainly by NMR and established structure (18). The tautomerism indicated in the formulae will be discussed in detail in Section 5.

Several reactions have been performed on the side chains of cercosporin. Oxidation (CrO_3 in pyridine) of the secondary alcohol groups of (18) and its dimethyl ether, respectively, afforded the corresponding mono- (19, 20) and diketones (21). Alkali-catalyzed retro-aldol reaction yields a derivative of 1,12-dimethylperylenequinone (22) (140). This

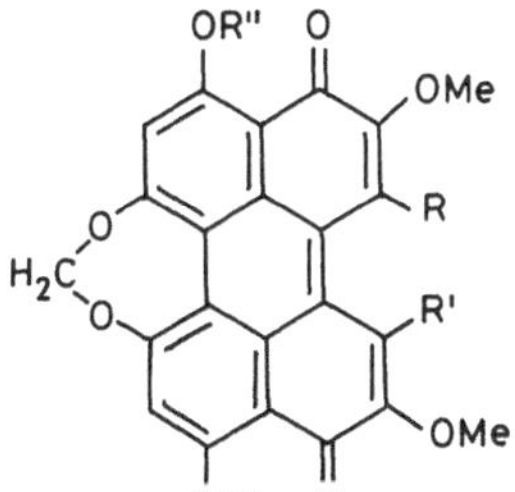

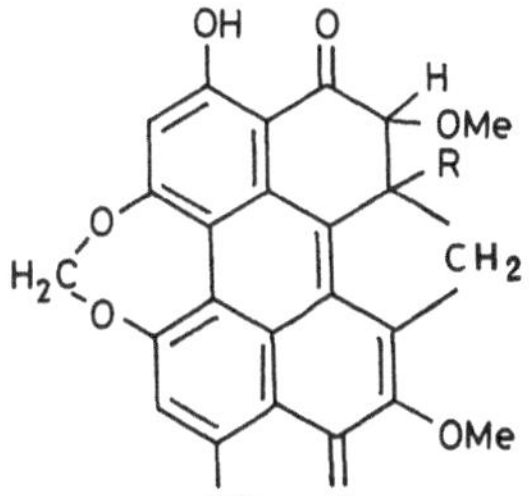

(19) R = CH₂COMe; R′ = CH₂CHOHMe; R″ = H
(20) R = CH₂COMe; R′ = CH₂CHOHMe; R″ = Me
(21) R = R′ = CH₂COMe; R″ = Me
(22) R = R′ = Me; R″ = H

(23) R = CH₂CHOHMe

(24)

cleavage may also involve only one of the two side chains and be accompanied by Michael attack of the methylene anion at C-1, the carbon β to the quinone carbonyl, to give the new ring-system (23) (108).

On the other hand, treatment with sulfuric acid induces cleavage of the methoxyls at C-2 and C-11, followed by cyclization of the side-chains, to form noranhydrocercosporin (24) (79, 108, 140).

The most significant transformation of cercosporin is the conversion of the dextrorotatory compound into a levorotatory isomer, isocerco-sporin, by warming in solution or in the solid state (78, 79). An alleged photochemical isomerization (96, 98) has not been confirmed (5, 77); exposure of the tetramethylether of cercosporin to sunlight gave tetra-methylnorcercosporin through cleavage of the methoxyls at positions 2 and 11 (77). Most physical (79) and chemical (108, 140) properties of the two isomers are very similar, but the compounds can be separated by chromatography. The isomerization is reversible, leading to an equi-librium mixture (ca. 1:1).

The intriguing nature of this isomerization became clear only after the structure of cercosporin had been elucidated: NMR and especially CD data showed that the two compounds are diastereoisomers and that this isomerism is caused by opposite chirality of their perylenequin-one chromophores which are forced out of coplanarity by the steric demands of the substituents at positions 1, 6, 7 and 12. The skewness of the chromophore of cercosporin has been confirmed by X-ray crys-tallography (102, 108). The fact that cercosporin and isocercosporin are diastereoisomers requires that the absolute configuration of the asymmetric carbons of the side chains be the same. It was established as R by application of the Horeau method (108).

So far only cercosporin has been observed as a natural product. The reason for this and the biosynthesis of the pigment in general are discussed in Section 6.

Esters of Cercosporin

One single species of the large form genus *Cercospora, C. setariae*, has been found to produce, besides cercosporin, several esters at the hydroxyalkyl groups: monoacetyl- (**25**), diacetyl- (**26**), monoacetyl-monobenzoyl- (**27**), and dibenzoylcercosporin (**28**). Their NMR spectra show a diagnostic downfield shift of the H-14 and/or H-17 protons with respect to the parent compound, those of the symmetrical esters showing only half of the signals as does cercosporin itself. Mild hydrolysis to cercosporin and comparison of the CD spectra established that the stereochemistry of these esters is identical with that of the parent compound (*7*). Compound (**27**) was obtained in crystals suitable for X-ray analysis (Fig. 3); it was therefore used to establish the absolute configuration of cercosporin (*108*).

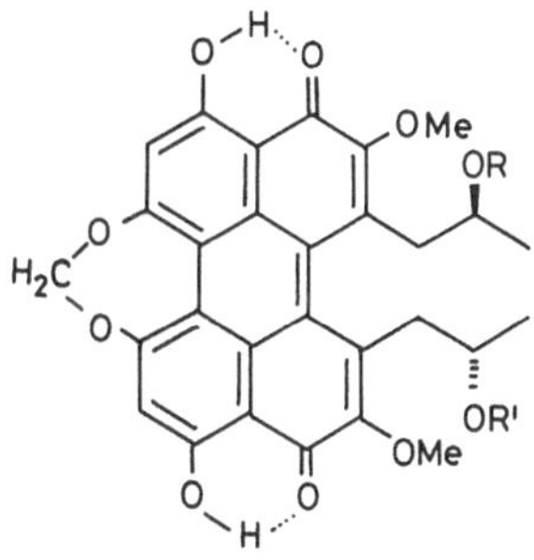

(**25**) R = COMe; R′ = H
(**26**) R = R′ = COMe
(**27**) R = COMe; R′ = COPh
(**28**) R = R′ = COPh

3.5 The Phleichromes

Phleichrome (**29**) is produced by the fungus *Cladosporium phlei*, a pathogen of timothy and other plants. Its structure was readily elucidated (*142*) by spectroscopic analysis and comparison with cercosporin, from which it differs only by the presence of two methoxy groups in positions 6 and 7 instead of the methylenedioxy group. Phleichrome is converted by heating in xylene into the diastereoisomer isophleichrome (**30**), the stereochemical relationship between the two compounds being the same as that between cercosporin and isocercosporin (*4*). Phleichrome, however, is a quasi-enantiomer of cercosporin, the configuration of the side chain asymmetric carbons being *S* (by

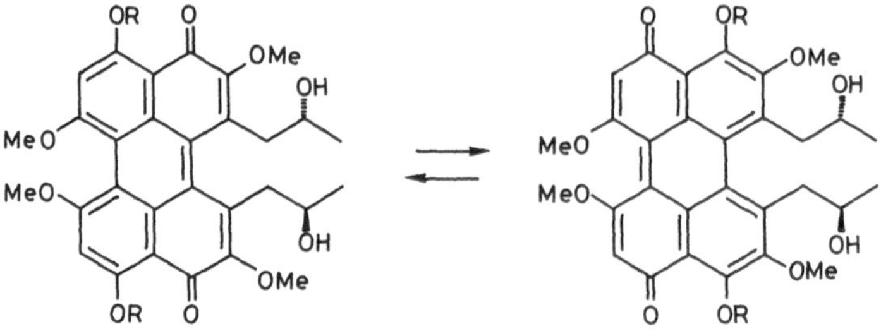

(29a, 30a) R = Me   (29, 30) R = H   (29b, 30b) R = Me

Horeau's method) and the axial chirality $P(S)$, as shown by comparison of the CD curves with those of cercosporin and isocercosporin (4) (Fig. 4 and 5 in Section 4).

The phenol-quinone tautomerism of both compounds (29) and (30) is shown by their NMR spectra (see Section 5.) and particularly by

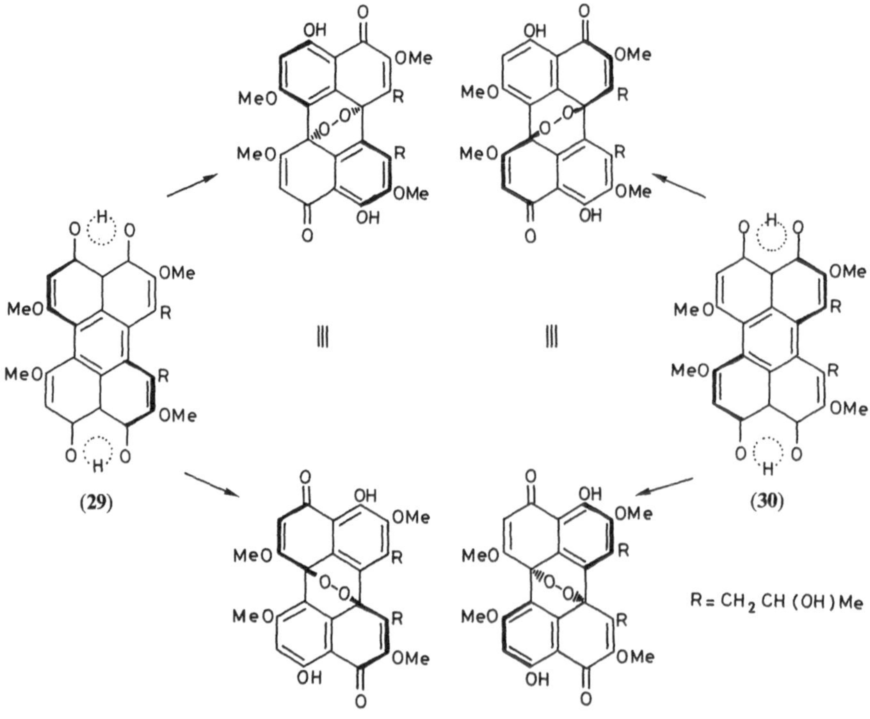

R = CH$_2$ CH (OH) Me

Scheme 2

the formation (5) from either of them of two isomeric dimethyl ethers derived from the tautomeric structures; cf. the analogous pairs of dimethyl ethers of elsinochrome A (Section 3.3.). Again, one member of both pairs is red (29a, 30a), the other one yellow (29b, 30b) (4). Here, again, the red color in these derivatives of the yellow perylene-3,10-quinone (2) is unexplained.

Photooxidation of phleichrome and isophleichrome in chloroform affords stereospecifically two endoperoxides, peroxyphleichrome (31) and peroxyisophleichrome (32), formed by the addition of one mole of dioxygen across the central ring of the reactive anthracene moiety of the two degenerate (equivalent) 3,9-dihydroxy-4,10-perylenequinone tautomers of (29) and (30) (Scheme 2) (4).
The stereospecificity of the reaction is a necessary consequence of the steric factors and symmetry properties exhibited by (29) and (30). The structure of the peroxides was established by ^{1}H and ^{13}C NMR spectroscopy. Assuming that the axial chirality of the peroxides is the same as that of the parent compounds (this assumption is also consistent with their CD spectra), the configuration of the new asymmetric centers C-6a and C-12a in peroxyphleichrome is S and R, respectively, and vice versa in the isomer (32) (4).

Recently, from a culture of *Cladosporium cladosporioides*, a compound indistinguishable from isophleichrome by TLC behavior and spectra was isolated (5). Examination of the CD absorption, however, established that the compound (33) is *ent*-isophleichrome.

3.6. The Cladochromes

In 1965 MAHADEVAN et al. (93) made the interesting observation that a red pigmentation is produced when the fungus *Cladosporium cucumerinum* is grown on etiolated seedlings of cucumbers (and also of other Cucurbitaceae (5)). Attempts (5, 111) to reproduce the pigmentation with normal seedlings or by growing the fungus on a variety of media, including juice pressed from etiolated cucumbers, have failed so far. The reasons for this observation are not yet clear: it can be speculated that only diseased seedlings contain active enzymes (phenoloxidases?) able to complete the biosynthesis of the red pigment or that normal seedlings produce phytoalexins capable of inhibiting some steps of the metabolism of the fungus. Extraction of the red seedlings afforded small amounts of at least five pigments. For only one of these, cladochrome A, an incomplete 4,9-dihydroxy-3,10-perylenequinone structure with four methoxy substituents and two 5-(2-hydroxy-4-oxohexyl) side chains was put forward by OVEREEM et al. (111), but

(33) R = R′ = H
(34) R = R′ = COCH₂CHOHMe
(35) R = COCH₂CHOHMe; R′ = COPh
(36) R = COPh, R′ = CO-p-PhOH

it was necessary to assume that the compound crystallized with two molecules of water in order to justify the analytical data which pointed to a formula $C_{38}H_{42}O_{14}$. Recent reexamination of the subject has shown that this is indeed the correct formula for cladochrome A (34) which is an ester of *ent*-isophleichrome with two moles of β-hydroxyiso-butyric acid (5). A second pigment, cladochrome B (35) was also isolated and found to be a mono-β-hydroxybutyrate monobenzoate of the same perylenequinone (5). These assignments are based on 300 MHz ^{1}H and ^{13}C NMR spectra and on mass spectra. Moreover, mild hydrolysis of the esters afforded *ent*-isophleichrome (33), thus establishing the stereochemistry of both compounds, except for the configuration of the β-hydroxybutyrate moiety.

A third member of the group, cladochrome C (36), again a diester of *ent*-isophleichrome, has been obtained from an *in vitro* culture of *Cladosporium cladosporioides* (5); in this case, tissue of the host plant was evidently *not* required for pigment formation.

3.7. Pigments from Mutants of *Cercospora kikuchii*

Irradiation of a culture of *Cercospora kikuchii* Matsumoto et To-moyasu (95) produces a mutant strain, first tentatively called *C. kikuchii* nov. var. (98), then *C. kikuchii* mut. *alba* (97). Besides cercosporin, a series of new metabolites (37–40) is produced by this strain. Precer-cosporin (37a), pseudocercosporin (37b), and protocercosporin (38) are 4,9-dihydroxy-3,10-perylenequinones and have structures which may be derived by further biosynthetic transformation of cercosporin. Precercosporin (37a) (97) is structurally identical with one of the prod-

(37a, b)   (38)

ucts (23) of alkali-catalyzed degradation of cercosporin (108). However, no information on the structural elucidation of the natural product (37a) is yet available and comparison of the two products has not been performed. Pseudocercosporin (37b) is claimed to be the diastereo-isomer of precercosporin, its relationship with the latter being the same as that of cercosporin and isocercosporin: they have opposite helicity of the chromophoric system (94). It is not known whether they are interconvertible. The structure of protocercosporin (38) is based on spectral data (101); the configuration is stated (97, 98) to have been established by X-ray analysis (no data published so far).

Puzzling from the viewpoint of biosynthesis are the structures pro-posed for two other pigments from the same source, neocercosporin (39) (98), (initially called neosporin (96)), and amphicercosporin (40) (101). Both compounds form reddish-violet (39) or reddish-purple (40) needles. In both structures phenolic hydroxyls occur in positions differ-ent from the usual ones, 4 and 9, in all other natural perylenequinones; pigment (39) is formulated as a 5,8-dihydroxyperylene-3,10-quinone, its congener (40) as the corresponding non-symmetrical 4,8-dihydroxy compound. These structural assignments are based to a large extent on spectroscopic data. The ^{1}H NMR spectrum of (39) clearly indicates a symmetrical structure, showing signals at δ 6.95 (2H, aromatic pro-tons) and 11.2 δ (2H, phenolic OH), while the spectrum of (40) contains pairs of signals at 7.10 and 7.12, and 11.80 and 14.60 δ (1H each); the signal from the methoxyls is likewise double (δ 4.28 and 4.30, 3H each; all data in CDCl$_3$). The rest of the spectra of both compounds resembles the one of cercosporin. In the IR spectra, bands at 1710 cm^{-1} for (39) and 1705 cm^{-1} for (40) are interpreted as being caused by the non-chelated quinone carbonyls; an additional band at 1630 cm^{-1} in the spectrum of (40) is ascribed to the chelated carbonyl.

However, the structures proposed for (39) and (40) do not explain some of the properties of these compounds. For instance, perylene-3,10-

(39) R=H; R'=OH
(40) R=OH; R'=H

quinone, which should be a good model for (39), forms yellow rather than violet crystals and shows its carbonyl band at 1650 cm^{-1} (25). Pigment (40) should resemble perylenequinones with one chelated quinone carbonyl, such as elsinochrome D (82) and the monomethyl ether of elsinochrome A (81) which are both orange and have broad carbonyl bands at ca. 1620–1630 cm^{-1} resulting from the overlap of the bands from the chelated and non-chelated carbonyl. The bands above 1700 cm^{-1} in the IR spectra of (39) and (40) thus remain unexplained.

The ORD, CD, and NMR spectra (94, 95) indicate that (37a), (38), and (40) belong stereochemically to the cercosporin series, whereas (37b) and (39) are similar to isocercosporin (see Section 4.).

Structures such as those given for (39) and (40) would require an unusual redox reaction or an NIH shift during their biosynthesis. Further detailed structural and biosynthetic studies on this group of compounds would thus be most welcome.

3.8. Hypocrellin

Hypocrellin (41), the red photodynamic pigment of the fungus *Hypocrella bambusae*, which grows in China on bamboo, represents a

(41)

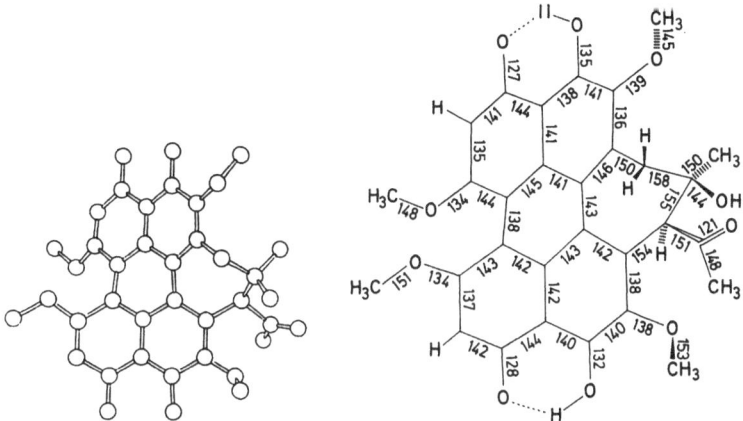

Fig. 2. X-ray structure, bond distances and relative configuration of hypocrellin (**41**).
(Reproduced with permission from ref. *135*)

further variant of the fungal perylenequinone structure, since here the
two three-carbon side chains are linked to form a seven-membered
ring. This peculiarity appeared from the ^{13}C NMR spectrum and was
confirmed by an X-ray analysis (*135*) (Fig. 2). This analysis also showed
the non-planarity of the perylenequinone system of hypocrellin, the
relative configuration of the asymmetric carbons of the alicyclic ring,
and the preference for a tautomeric structure in the solid state with
a quinonoid "left" moiety.

Comparison of the CD curves of hypocrellin with those of the
cercosporin group led to the hypothesis that hypocrellin may possess
the same axial chirality as cercosporin (*103*), although there are large
differences in intensity of the bands (Fig. 4 in Section 4.). If this hypothesis is correct, the axial chirality is known and, together with X-ray

analysis, requires that carbons 14 and 15 of hypocrellin must have configuration R and S, respectively.

Hypocrellin was the first of the perylenequinone pigments shown to form an endoperoxide by photooxidation. To peroxyhypocrellin the structure (42) was assigned on the basis of a reasonable mechanism and of the ^{13}C NMR spectrum (see, however, the footnote on page 1390 of ref. (4)).

Since, however, hypocrellin lacks a C_2 axis of symmetry, an alternative structure with the peroxy bridge linking carbons 6a and 12a is equally possible.

3.9. The Aphins

The group (Group C) of perylenequinones obtainable from aphids consists of three pigments: the stereoisomeric erythroaphins *fb* (4) and *sl* (5), and rhodoaphin-*be* (6), the dihydroxy derivative of a third stereoisomer, erythroaphin-*tt* (43) (18).

The classical research of Lord TODD and his associates, documented in a long series of papers from 1948 on, led to detailed understanding of the structure and stereochemistry of these complex substances and of the formation of (4) and (5). Of these compounds, only the last one, rhodoaphin-*be* (6), can actually be considered a genuine natural product. It occurs (89) as the aglycon of a little investigated glycoside heteroaphin in certain scarce primitive aphid species of the genus *Hormaphis* (formerly called *Hamamelistes*). It is (18) the dihydroxy derivative of erythroaphin-*tt*, which has never been obtained from any natural source but can be made by catalytic hydrogenolysis of rhodoaphin-*be* (18) or via photochemical epimerization of the leucotetraacetates of (4) and (5) (37).

Erythroaphin-*fb* (4) R = R′ = αH
Erythroaphin-*sl* (5) R = βH; R′ = αH
Erythroaphin-*tt* (43) R = R′ = βH
Rhodoaphin-*be* (6) R = R′ = βOH

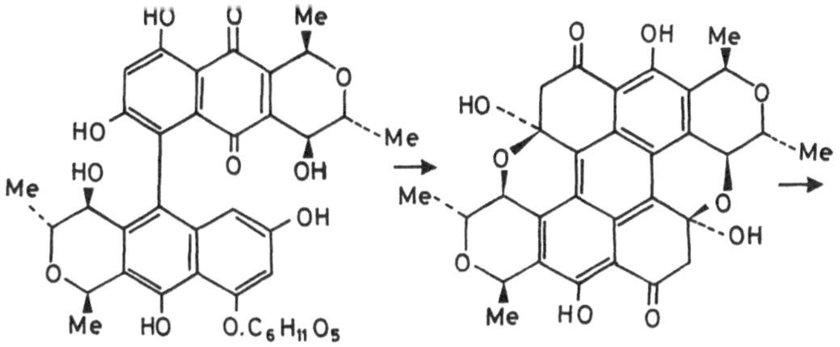

Scheme 3. Formation of erythroaphin-*fb* (4) from protoaphin-*fb*

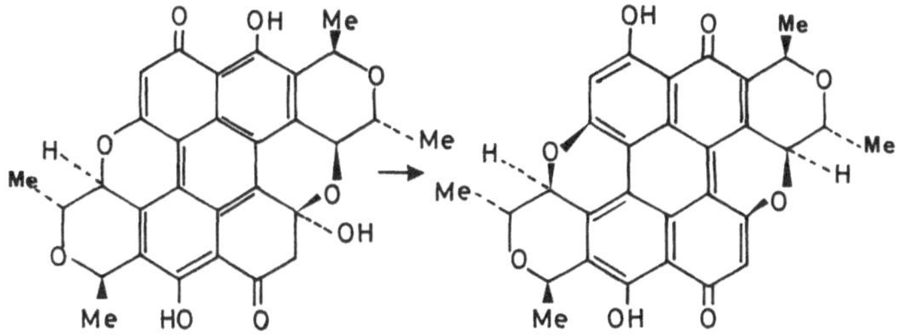

The two remaining perylenequinones, the erythroaphins-*fb* (4) and *sl* (5), have been isolated from many species of aphids but by no means from all. They do, however, not exist as such in these insects, but are formed enzymatically after the death of the aphids from their native pigments (*40*), the stereoisomeric protoaphins-*fb* (44) and *sl* (55). This conversion proceeds through two consecutive intermediates, the xanthoaphins and chrysoaphins (Scheme 3).

The extensive work of the Cambridge School of TODD *et al.* has been reviewed several times (*26, 128, 129, 130*); these papers should be consulted for details which cannot be given within the space limitations of this review.

Since neither the erythroaphins nor their precursors, the xantho- and chrysoaphins, occur as such in nature but are products of post-mortem transformations, it may seem debatable whether they should be included in a review on the naturally occurring perylenequinones. However, intrinsic interest of their chemistry, formation and transformations would alone justify their inclusion; even more important is the fact that they were the first perylenequinones obtainable from natural sources, even if they do not exist there as such, which were studied exhaustively and that this seminal study has provided much of the methodology used subsequently in this field. It would be difficult indeed to give an account of the work on, e.g., the elsinochromes, if an account of work on the aphid pigments were lacking.

The suffixes *fb*, *sl*, *be* etc. were introduced early (*70*) by TODD *et al.* to indicate
the aphid species from which a particular compound was first obtained. After the elucidation of structure and stereochemistry, they now serve as convenient indicators of the
configurational affinities.

The conversion of the protoaphins into the erythroaphins is initiated
by enzymatic cleavage of the glycosidic bond. The resulting epimeric
aglycones are transformed into the fully aromatic pigments in three
distinct steps: formation of the very unstable, yellow xanthoaphins,
conversion of these into the somewhat more stable, orange chrysoaphins by elimination of one molecule of water, and finally, loss of
another molecule of water from these to give the erythroaphins. The
xantho- and chrysoaphins contain the perylene skeleton in partially
reduced form. They are dealt with here, rather than in Section 8., for
the sake of clarity. Several stereoisomers of both types of intermediate
are known. Some of the steps of this sequence are catalyzed by enzymes,
or by acid, the elimination of water from the xantho- and chrysoaphins
also by base (*29*).

The structures of the compounds of this sequence are given in
Scheme 3. In this Scheme, the substances of the *fb* series are shown,
since here only single stereoisomers of xantho- and chrysoaphin are
involved. This is a consequence of the fact that in this series the two
nonaromatic parts of the molecules have identical stereochemistry. In
the *sl*-series, where these moieties have different configurations, two
stereoisomeric xanthoaphins and three stereoisomeric chrysoaphins
have been isolated.

Actually the configurational identity of the two non-aromatic moieties in the *fb* series
is not sufficient in itself to account for formation of only one xanthoaphin-*fb*. There
is much spectroscopic evidence (to be discussed below) for the noncoplanarity of the
two naphthalene systems in the protoaphins. The formation of the more nearly planar
ring system of the xanthoaphins must therefore involve rotation, either clockwise or coun-
terclockwise, of one of these moieties with respect to the other. As a consequence, two
different, stereoisomeric xanthoaphins could result even in the *fb* series. However, one
of these rotations would produce a highly strained molecule and is thus very unlikely
to occur, while the opposite mode of rotation (indicated in Scheme 13, Section 6.) yields
favorable stereochemistry at the two newly formed asymmetric centers.

3.9.1. The Protoaphins

These stereoisomeric pigments, (**44**) and (**55**), are yellow, non fluorescent crystalline β-D-glycopyranosides (*33*, *35*). Since they undergo
very rapid enzymatic transformations after death of the insects, they
can be isolated only if these enzymes are first inactivated by killing
the aphids through treatment with water at 70 °C (*22*). Most aphid

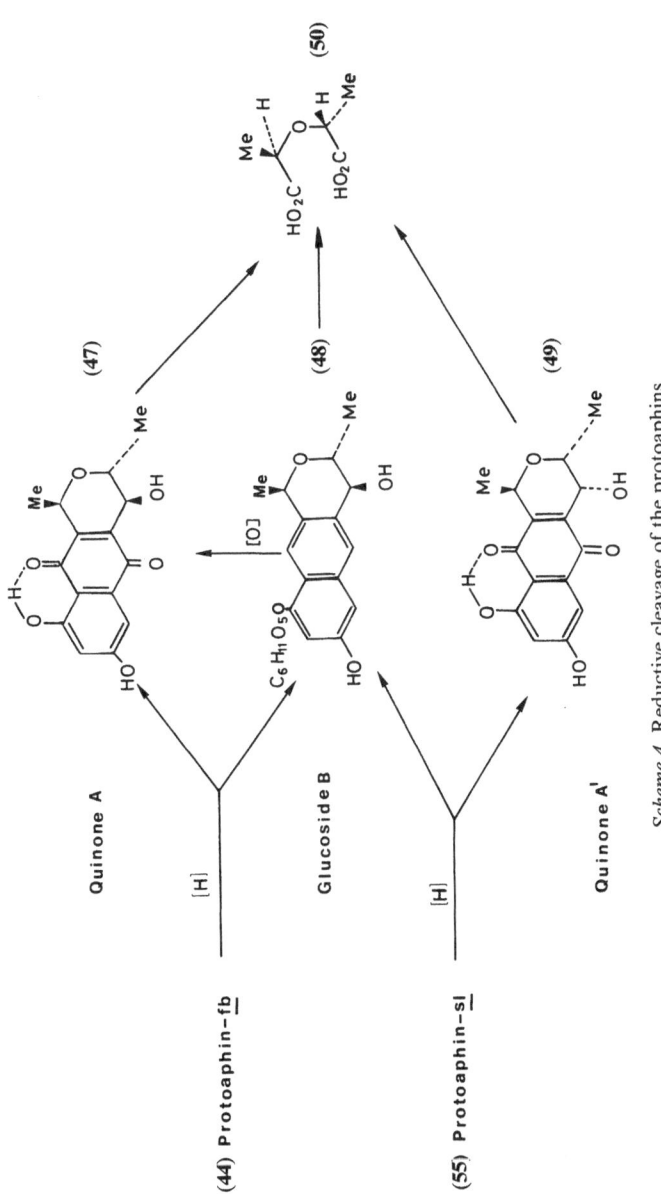

Scheme 4. Reductive cleavage of the protoaphins

species which produce these pigments contain only one or the other stereoisomer, but cases of co-occurence of both have been observed in some species, even in the same individual (*11*).

Hydrolysis of protoaphin-*fb* under nitrogen yields the aglycone which is very sensitive to oxygen but otherwise quite stable. The cleavage can be acid catalyzed, or enzymatic in presence of boric acid (*33*), which keeps the aglycone as a borate complex.

The protoaphins show the usual redox behavior of quinones (*40*, *22*). Their leuco-dodecaacetates have absorption spectra of naphthalenic type (*40*). Elucidation of structure and complete stereochemistry (both relative and absolute) of the protoaphins was achieved largely through study (*31*, *40*) of their reductive cleavage.

Reduction with neutral dithionite cleaves the central bond of both protoaphins (*40*), yielding two fragments, the one an orange quinone, the other a colorless glycoside. The latter, called glycoside B (**48**), is the same whether derived from protoaphin-*fb* (**44**) or *sl* (**55**); in contrast, the two quinones are epimeric, protoaphin-*fb* yielding quinone A (**47**), the *sl* stereomer giving quinone A′ (**49**). The structures of these three products follow from spectroscopic data, from oxidation of glycoside B to quinone A by Frémy's salt, and from the finding that all three compounds are oxidized to *D,D*-(+)-dilactic acid (**50**) of known absolute configuration (*64*) (Scheme 4).

These facts, together with consideration of the transformation of the protoaphins into derivatives of perylene, permitted elucidation of structure and stereochemistry of the protoaphins and the stereochemistry of the pigments derived from them. The results show that the absolute configuration of the methyl-bearing carbons is always the same and that the difference between the compounds of the *fb* and *sl* series rests on epimerism of the benzylic hydroxyl in quinones A and A′. The asymmetry at this center is destroyed during the oxidation to dilactic acids; its configuration in quinones A and A′ was proved by study of the NMR signals of the pair of protons on this and the adjacent carbon. Together, these studies have played a major role in establishing the structures and absolute configurations of the xantho-, chryso-, and erythroaphins.

The cleavage products of the protoaphins can be recombined with remarkable ease (aqueous solution buffered to pH 6.6, room temperature) to give almost quantitatively a mixture of two isomeric coupling products, of which one – the minor one – is protoaphin-*fb* or *sl*, respectively; the major isomer is formed through coupling of the resorcinol rings of the components (*31*). Theoretically, the two moieties, quinone A (or A′) and glycoside B, can be combined in several different ways, but of the resulting structures only expressions (**44**) and (**55**) for the two protoaphins are compatible with the subsequent transformation into the erythroaphins-*fb* (**4**) and *sl* (**5**), respectively.

The absorption spectra of the protoaphins are very similar to the curve resulting from summation of the spectra of their cleavage products (*40*). This fact, together with the naphthalenic spectrum of the leuco-dodecaacetates, already mentioned, shows the absence of any resonance interaction between the two bicyclic moieties which can thus not be coplanar.

Many aphid species contain an additional compound resembling the protoaphins (*11*). This pigment, deoxyprotoaphin, was identified as (**51**) by hydrogenolysis to glycoside B and deoxyquinone A, previously obtained (*19*) through reduction of quinone A with sodium stannite. Surprisingly, deoxyprotoaphin also occurs as a minor pigment in aphids of the genus *Dactynotus*; insects of this genus otherwise produce pigments different from the aphins discussed in this review. Like these, the pigments from *Dactynotus* spp are built up from two *lin*-naphthopyran units; here, however, these units are held together by two carbon-oxygen bonds instead of the carbon-carbon bonds which provide the linkage in the aphins. Extracts of *A. fabae* convert deoxyprotoaphin (**51**) into an analog (**52**) of chrysoaphin which is dehydrated to a perylenequinone (**53**) by hydrochloric acid (Scheme 5).

Other pigments resembling the protoaphins have been observed in many aphid species (*11*); they have not been described in detail.

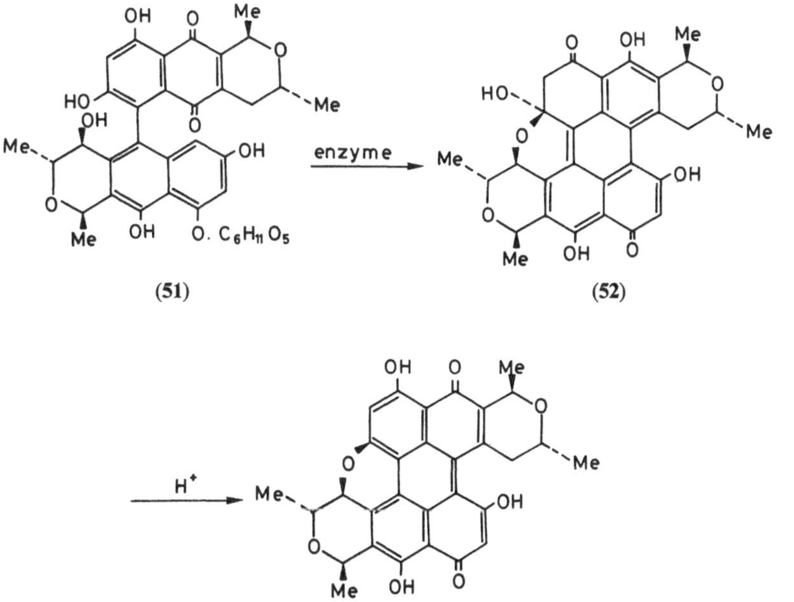

Scheme 5. Transformation of deoxyprotoaphin (**51**) to a perylenequinone (**53**)

3.9.2. The Xanthoaphins and the Chrysoaphins

These very labile, highly fluorescent pigments are intermediates in the formation of the erythroaphins from the protoaphins (see Scheme 3.). Several stereoisomers of either group are known, *viz.* three xanthoaphins and four chrysoaphins. In the formation of the xanthoaphins the perylene skeleton is generated, together with two hemiketal functions which constitute new centers of asymmetry. In the next step which yields the chrysoaphins one of these functions disappears while the other one survives, to be removed in the last step of the sequence. The presence of these hemiketal groupings accounts for the great lability of the pigments and for the remarkable ease with which they epimerize, with the compounds of the *fb*-series being the most stable ones. As expected, the xanthoaphins with their two hemiketal functions are even more sensitive than the chrysoaphins, and the lability is especially great in those compounds (xanthoaphin-*sl*-2 and chrysoaphin-*sl*-3) in which one of the oxygen-bridges has *trans*-stereochemistry (Scheme 6).

Because of this excessive sensitivity the structure and stereochemistry of the xanthoaphins and chrysoaphins had to be derived to a large extent from consideration of those of the protoaphins and erythroaphins and from spectroscopic comparison with model compounds.

The xanthoaphins do not give the characteristic redox reactions of quinones. The presence of the ring system of anthracene was suggested initially (*29*) by their absorption spectra, that of non-chelated hydroxyl and of chelated keto groups by the infrared bands at 3300 and 1625 cm^{-1}, respectively. Proof for the actual structure of the chromophoric part of the molecule comes from the close resemblance of the spectrum of xanthoaphin in the visible region to that of 1,5-diacetyl-2,6-dihydroanthracene (**54**) (*41*).

It has already been mentioned that the conversion of protoaphin-*fb* into the corresponding xanthoaphin yields one single stereoisomer; the epimer, which could result from an alternative mode of cyclization,

(**45**)   (**54**)

would apparently be prohibitively strained. These stereochemical constraints must be less rigid in the *sl*-series. Here, in addition to the xanthoaphin-*sl* initially obtained (*29*, *58*) (now renamed xanthoaphin-*sl*-1 (**56**)), an epimer, xanthoaphin-*sl*-2 (**57**) has been isolated (*12*), which must be formed through this alternative cyclization. Models of this xanthoaphin-*sl*-2 can be built only with some distortion of the anthracene. As expected from this, the substance is extremely unstable; on standing at room temperature, either as the solid or in aqueous suspension, it is converted into a mixture of xanthoaphin-*sl*-1 (**56**) and chrysoaphin-*sl*-1 (**59**). Surprisingly, xanthoaphin-*sl*-2 (**57**) is formed exclusively on very brief treatment of protoaphin-*sl* with extracts from the aphid *Tuberolachnus salignus*, the chief source of pigments of the *sl*-series. In contrast, even the briefest exposure of protoaphin-*sl* (**55**) to extracts from *Aphis fabae* yielded only xanthoaphin-*sl*-1 (**56**).

The structures, configurations and relationships of the various stereoisomers of the xanthoaphins-*sl* and chrysoaphins-*sl* are shown in Scheme 6.

Elimination of the elements of water from one of the hemiketal groupings of the xanthoaphins produces the orange highly fluorescent quinonoid chrysoaphins. Their overall structure follows from their formation, their conversion into the erythroaphins, from spectroscopic evidence, and from study of a model compound. This substance (**61**), was obtained (*31*) by coupling of quinone A (**47**) with 1,8-dihydroxynaphthalene in a reaction analogous to the reconstitution of the protoaphins from their cleavage products and heating the resulting binaphthyl derivative in aqueous solution. The absorption spectrum of the orange (**61**) so obtained closely resembles those of the chrysoaphins.

The elimination of water which transforms the xanthoaphins into the chrysoaphins can yield only a single product chrysoaphin-*fb* (**46**) from the structurally and stereochemically symmetrical xanthoaphin-*fb* (**45**).

(**61**)

U. Weiss, L. Merlini, and G. Nasini:

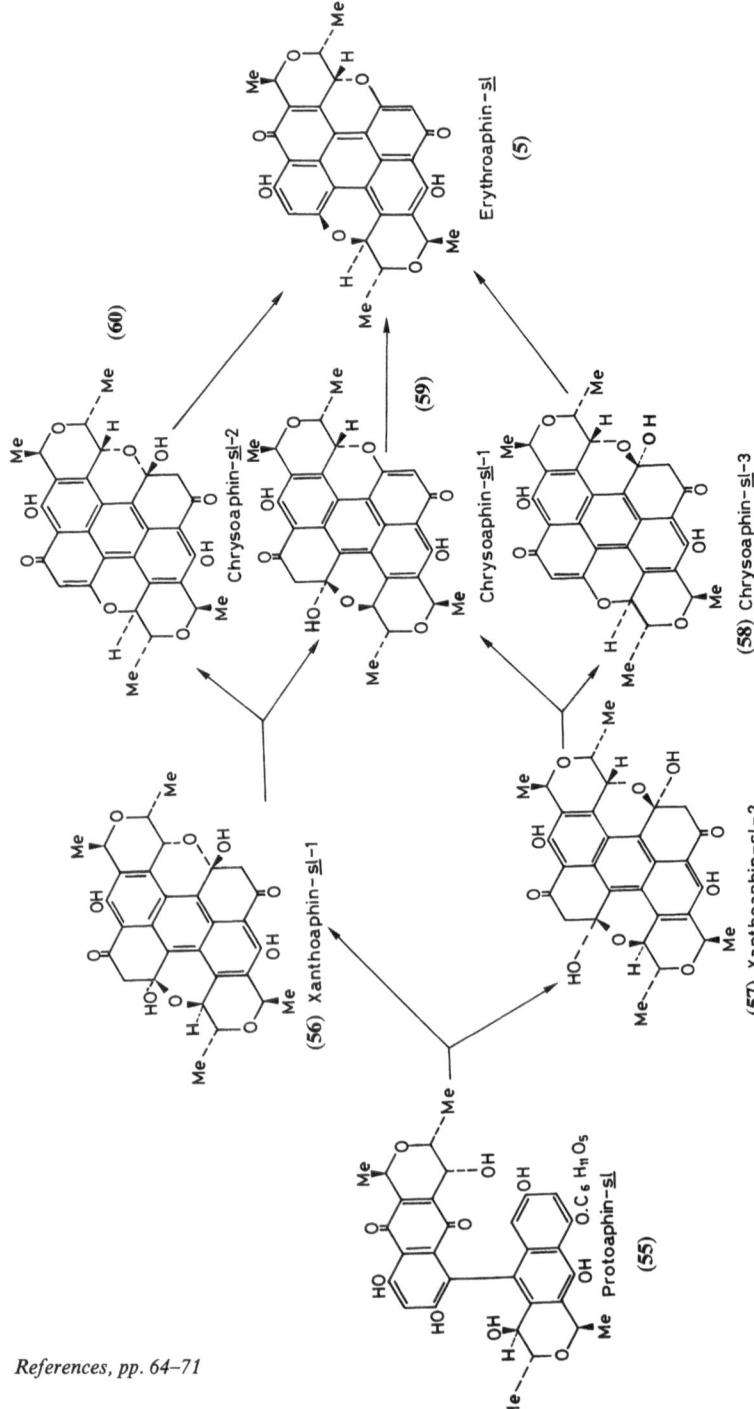

Scheme 6. Transformation of protoaphin-*sl* into the corresponding xantho-, chryso-, and erythroaphins

In the *sl* series, however, the two sides of both xanthoaphin-*sl*-1 (**56**) and *sl*-2 (**57**) are stereochemically different from each other and it matters which side suffers loss of water. Altogether, three stereoisomeric chrysoaphins-*sl* are possible and known. Of these, the *trans*-compound chrysoaphin-*sl*-3 (**60**) is again highly unstable, although not quite as much so as xanthoaphin-*sl*-2 (**57**). All these chrysoaphins readily lose one molecule of water to give erythroaphin-*sl* (**5**), the stable final member of the series. Scheme 6 shows the structure, configuration, and relationships of the compounds of the *sl*-series.

The enzymes involved in the first two steps of the conversion of the protoaphins into the erythroaphins have been studied. The protoaphins are hydrolyzed by extracts from aphids, even by those from species which do not yield aphins, e.g. *D. jaceae*. In the latter case the reaction stops after the cleavage of the glycosidic bond. In contrast the resulting aglycone is converted further into xanthoaphin by extracts from aphin producing species. It has already been mentioned that the aglycone of protoaphin-*sl* is cyclized initially to xanthoaphin-*sl*-2 by extracts from *T. salignus*, but into xanthoaphin-*sl*-1 by those from *A. fabae* (*20*).

Although the protoaphins are β-glycosides, they are resistant to emulsin, the typical β-glycosidase; apparently, the enzyme is inhibited or inactivated by the quinonoid moiety of the substrate (*35*). In contrast, glycoside B (**48**) is cleaved smoothly by emulsin, if further oxidation of the resulting aglycone is prevented by working under nitrogen.

No attempt at purification of the glycosidases from aphids seems to have been made. In contrast, the enzyme from the aphid *Eriosoma lanigerum* which converts the aglycone of protoaphin-*fb* into xanthoaphin-*fb* has been purified almost to homogeneity by chromatography on Sephadex (*44*). This enzyme has a molecular weight of 120000 ± 2000 dalton and contains 14–15 glucosamine residues per mole of protein.

3.9.3. The Erythroaphins; Rhodoaphin

Both erythroaphins (**4** and **5**), $C_{30}H_{22}O_8$, are crystalline red pigments which give strongly fluorescent solutions in organic solvents. The near-identity of their absorption spectra and the parallelism of their chemical transformations shows them to be stereoisomers. Their quinonoid nature is proved by the usual redox reactions (e.g. the change from green to pink on addition of dithionite to the alkaline solution), reversed by aeration. The presence of a carbonyl band at 1623 cm^{-1} together with the absence of a hydroxyl stretching band indicate a completely chelated extended quinone system as does the green color of the alkaline solutions. The chromophore is identified unequivocally as a 4,9-dihydroxyperylene-3,10-quinone by the close agreement (30) of the electronic spectra of the erythroaphins themselves and their yellow diacetates and leucotetracetates with those of 4,9-dihydroxyperylene-3,10-quinone and its corresponding derivatives (see Fig. 1). The formation of mellitic acid on vigorous oxidation with nitric acid is in agreement with this identification, which accounts for twenty carbon and four oxygen atoms of the two pigments.

Fusion with zinc dust produces (24) perylene, 1,12-benzoperylene, and coronene, indicating the presence of carbon substituents in positions adjacent to the ring junctions.

The relatively simple NMR spectrum of erythroaphin-fb[1] (4) shows a symmetrical structure; it also establishes the presence of two protons on the chromophore, and of two different pairs of secondary methyls (12).

The complete structure and stereochemistry (both relative and absolute) of the erythroaphins follow from these data and from the work on the reductive cleavage of the parent protoaphins; this work has been discussed in Section 3.9.1. and is shown there diagrammatically in Scheme 4. That the formation of the erythroaphins from the protoaphins does not involve changes in structure and stereochemistry of their nonaromatic moieties is shown by the analogy of the NMR spectra of both groups of pigments (38, 40) and especially by the partial synthesis of erythroaphin-fb through oxidative coupling (neutral ferricyanide) of two molecules of glycoside B (48), which yields a diglycoside readily transformed into erythroaphin-fb by dilute hydrochloric acid (32). (The intermediate diglycoside is actually derived from the 3,9-quinonoid tautomer of the erythroaphin; see Scheme 8).

Structure and stereochemistry of rhodoaphin-be (6) follow from the simplicity of its ^{1}H NMR spectrum, indicating a symmetrical molecule, and from interrelations with other aphin compounds: hydrogenolysis mostly to erythroaphin-tt (43), (a reaction shown not to alter the configuration) and acid-catalyzed equilibration to the known dihydroxy derivative of erythroaphin-fb (18).

A great wealth of information on the chemical properties and transformations of the erythroaphins has been accumulated in the laboratory of Lord Todd; only a small part of it can be mentioned here.

Some unusual reactions of these pigments have been found to take place at their benzylic carbons, in contrast to the halogenations, which proceed normally with replacement of the two hydrogens on the chromophore (23, 38).

---

[1] The greater complexity of the spectrum of the sl-isomer is caused by the stereochemical inequality of the two non-aromatic moieties of the molecule.

Addition of amines (23) yields stable amino analogs of ketals; analogous derivatives are also formed from the dibromoerythroaphins with retention of the two halogens. The reaction is interpreted as proceeding via tautomeric forms in which the relevant carbons are doubly bonded. In agreement with this view, amination of compounds of the sl series yields diamino derivatives of erythroaphin fb which are more stable than their epimers and are hence often formed preferentially during reactions which can proceed with stereochemical epimerizations. Many such reactions have been observed.

This novel amination at the benzylic position of a quinone, rather than in the usual fashion on the ring, also takes place in simple quinones such as duroquinone (38). Thiele acetoxylation proceeds in the same way as the amination and the benzylic positions can also be hydroxylated by periodate (36, 75).

It has been pointed out in the introduction that the structure of the erythroaphins, as members of Group C, are centrosymmetric (C_{2h}) (disregarding complications based on different configurations of the two non-chromophoric moieties), while those of the mold pigments of group B have C_{2v} symmetry.

The tautomerism of the aphins is discussed in Section 5. Because of the different symmetry the situation is markedly different from that found in compounds of group B (e.g. elsinochrome A) with its two isomeric dimethyl ethers both derived from the 3,10-quinone. Only one such ether is possible in the case of erythroaphin-fb, but this compound likewise forms two methyl ethers; one of these, however, is a 3,9-quinone (see formulae C and D in Scheme 8 of Section 5.).

The formation of the erythroaphins from their precursors is well understood (it could hardly be called a biosynthesis). In contrast, the biosynthesis of rhodoaphin-be has apparently not been studied in spite of the fact that this pigment is the only true naturally occurring aphin. The relative unavailability of the rhodoaphin producing aphid species seems to be responsible for this gap. No precursors analogous to the proto-, xantho- or chrysoaphins have been reported. A chemically plausible hypothesis for the biosynthesis of rhodoaphin is given in the Section on Biosynthesis (Section 6.), which also includes a discussion of the puzzling ultimate origin of these polyketides occurring in animals.

4. Stereochemistry

4,9-Dihydroxy-3,10-perylenequinone (3) is an achiral planar substance. The presence of side chains with asymmetric carbon atoms
may give rise to stereoisomers.

In most perylenequinones of fungal origin, structural elements such
as the two bulky methoxy groups or a strained seven-membered ring
in positions 6 and 7, together with the two C_3 side chains in positions
1 and 12, are the source of enough steric hindrance to force the penta-
cyclic perylenequinone system into a nonplanar helical shape. This helicity generates an element of asymmetry, the axial chirality (28) of the
molecule.

If the nonchromophoric part of the molecule contains asymmetric
carbons, diastereoisomerism may arise.

The early (78) observation of thermal isomerization of cercosporin
(18) to the diastereoisomer isocercosporin (62) was correctly explained
(140) by inversion of the ring helix (Scheme 7). Since this process inter-
converts two diastereoisomers, the two asymmetric carbon atoms of
the side chain of cercosporin and isocercosporin must necessarily have
the same configuration (140). These conclusions, supported by CD
and ORD evidence (140), were confirmed by oxidation of cercosporin
and isocercosporin dimethyl ethers, respectively, to the antipodal ke-

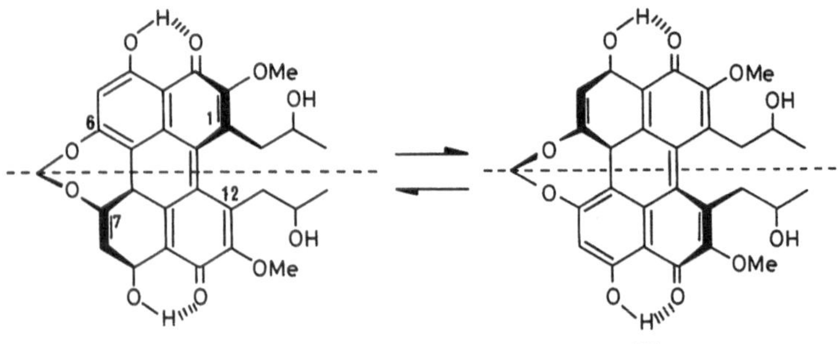

Scheme 7. Noncoplanarity of the perylenequinone system in cercosporin (18) and isocercosporin (62). In the two lower diagrams, the formulae are viewed edgewise towards
C-9 and C-10. (Redrawn, with permission, from ref. 140)

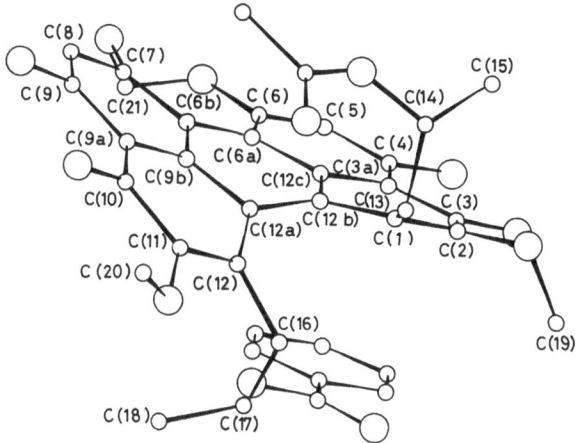

Fig. 3. X-ray structure of cercosporin monoacetate monobenzoate (**27**). (Reprinted with permission from ref. *108*)

tones (**21**) (*108*) which were still strongly optically active, though without asymmetric carbons, by the assignment of the absolute configuration (*R*) at C-14 and C-17 through Horeau's method of kinetic resolution (*108*), and by X-ray analysis of cercosporin itself (*102*) and of one of its derivatives (*108*) (Fig. 3). Oxygen anomalous dispersion measurement (*108*) gave the sign of the axial chirality, which is *M* (*R*) (**28**) for cercosporin.

The same behavior is shown by phleichrome (**29**), which equilibrates (*4*) with isophleichrome (**30**); here, however, the natural isomer has *P* axial chirality and *S* configuration of the side chain carbons, as shown by comparison of its CD curve with that of cercosporin (Fig. 4 and 5), and again also *via* Horeau's method (*4*). Cladochromes A (**34**), B (**35**), and C (**36**) were correlated with isophleichrome (**30**) by hydrolysis of the ester groups in the side chains to *ent*-isophleichrome (**33**), and by CD measurements (*5*).

Scarcity of published data, especially on the possible isomerization of the compounds, allows only an incomplete description of the stereochemistry of the group of substances isolated from *Cercospora kikuchii* mut. *alba*. From ORD data protocercosporin (**38**), amphicercosporin (**40**) and precercosporin (**37a**) seem to possess *M* axial chirality (*94*), opposite to that of neocercosporin (**39**) (*98*) and pseudocercosporin (**37b**) (*94*). Amphicercosporin is reported to isomerize photochemically or by treatment with polar solvents, but no structure is given for the resulting product, whereas neocercosporin is reported not to be photo-

U. WEISS, L. MERLINI, and G. NASINI:

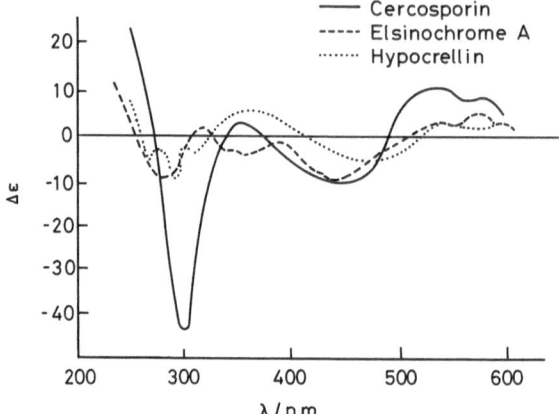

Fig. 4. Circular dichroism spectra of cercosporin (**18**), elsinochrome A (**14**), and hypocrellin (**41**)

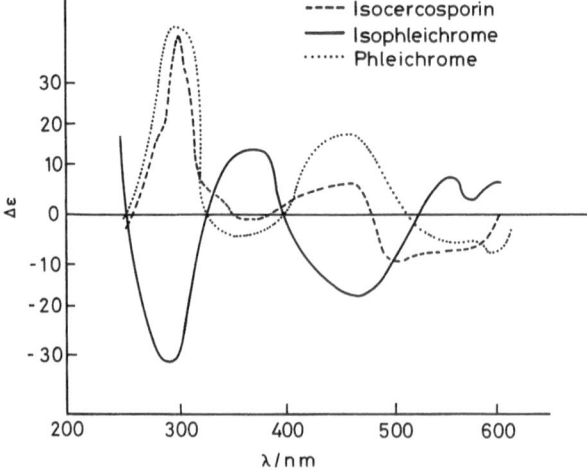

Fig. 5. Circular dichroism spectra of isocercosporin (**62**), isophleichrome (**30**) and phleichrome (**29**)

sensitive (*98*). The relative configuration of protocercosporin results from X-ray analysis (*101*, no experimental details given).

The presence of flexible side chains on the perylenequinone rings of cercosporin, phleichrome, and their congeners raises the problem of possible preferred conformations of these chains. Again, cercosporin and isocercosporin will be discussed in detail, whereas the results ob-

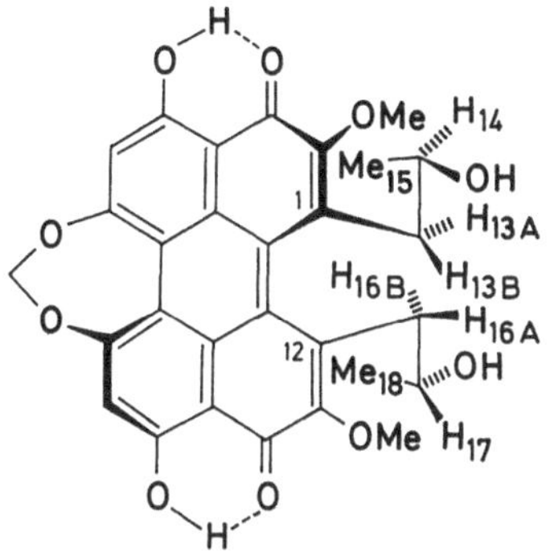

Fig. 6. Preferred conformation of cercosporin (**18**) and isocercosporin (**62**)

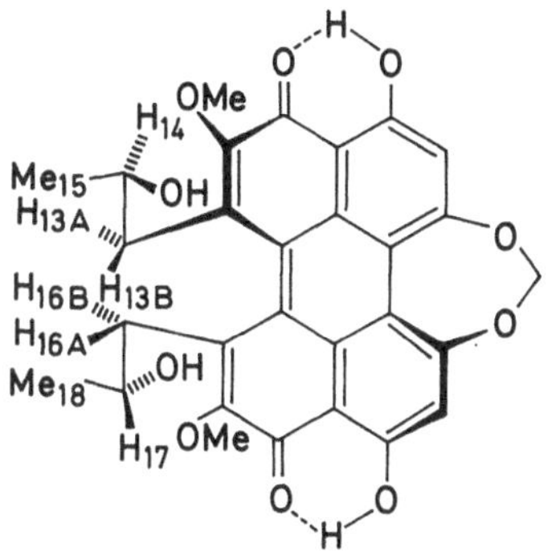

tained for other members of the group shall be summarized. The structures of cercosporin and isocercosporin contain a C_2 axis; consequently, the corresponding atoms in the two halves of the molecules
give identical signals in the NMR spectra. That in solution the side
chains possess different conformations in the two diastereoisomers was
initially suggested by the finding (*140*) that the chemical shifts of the
signals from the pair of methyls, C-15 and C-18, of cercosporin and
its derivatives (esters, ethers) are different from those of isocercosporin
and its derivatives. The same goes for the coupling constants of the
protons -C$\underline{H}_2$-C$\underline{H}$-OH-, in the side chains.

The chemical shift of Me-15 and Me-18 (δ 0.60 in CDCl$_3$) of cercosporin indicates that these groups lie above the perylenequinone system, enough so to undergo a shielding effect, whereas in isocercosporin
(δ 0.96) they are outside of the area of the ring current. Complete
analysis of the ^{13}C NMR spectra (*5*) provided coupling constants (^{2}J
and ^{3}J) between C-1 and C-2 and both the methylene protons H-13 A
and H-13 B; from these the orientation of the methylene protons with
respect to the aromatic system could be obtained. Studies of nuclear
Overhauser effects gave information on the rotation around the bond
between C-13 and C-14. Combination of all these data for both compounds led to the preferred conformations for (**18**) and (**62**) shown
in Fig. 6. For the sake of brevity, a "cercosporin-like" conformation
as in Fig. 6 will be called "OH-exo" and the corresponding "isocercosporin-like" one (Fig. 6) "OH-endo". This indicates the orientation
of the OH groups in both side chains with reference to the perylenequinone system. In both cases one of the side chains is bent upwards,
the other downwards with respect to the chromophore (see also Fig. 3).

Similar studies of ^{1}H and ^{13}C NMR spectra of phleichrome and isophleichrome permit the conclusion that the preferred conformations of the side chains of this pair are entirely similar to those of cercosporin and isocercosporin, respectively.

The thermal inversion of the helix in phleichrome and cercosporin is accompanied by a change in the conformations (OH-exo) of the side chains which assume a different preferred orientation (OH-endo) (Fig. 6). The process leading to this new conformation can be viewed as follows: the inversion of the helix would force the two side chains towards each other; they must therefore undergo a conrotatory movement to put themselves again on opposite sides of the main ring system. However, because of the chirality of the CHOH carbon, simple rotation of 180° around the bonds linking C-1 and C-13 (and C-12 and C-16), would give rise to a situation where the two methyl groups of the chain protrude toward the OCH$_3$ substituents. It can then be envisaged that a further rotation around the C-14/C-15 (resp. C-17/C-18) bonds relieves this strain, yielding a more favorable conformation of the type shown in Fig. 6.

It is understandable that such a process which involves closely spaced groups of atoms must have a fairly high barrier of inversion, so that equilibration of both pairs of isomers is slow even at elevated temperatures. That the interaction of the side chains plays a determining role in this respect is shown by the fact that this interaction is reduced by cyclization to noranhydrocercosporin (**24**). The same antipode of this evidently planar compound is obtained from cercosporin and isocercosporin, with the only contribution to the CD absorption made by the asymmetric carbons (*108*). On the other hand, the easy racemization at room temperature of helical dibenzo- and dinaphtho-dioxepin is well known (*67*). The magnitude of the inversion barrier for cercosporin was estimated by measuring the rate constant of the mutarotation at different temperatures. The values of $\Delta H^{++} = 20.0$ kcal·mole^{-1} and $\Delta S^{++} = 20.8$ e.u. were obtained (*108*).

The configuration and the preferred conformation of the clado-chromes were established by a process similar to that described for phleichrome (*5*).

Inspection of the configuration and behavior of the different pairs of isomers leads to the conclusion that the choice of a OH-exo or OH-endo conformation is regulated by *both* the sense of axial chirality and the configuration of the side-chain carbons (see Section 6. for the biogenetic implications).

In the elsinochromes, the carbons (C-1 and C-2) corresponding to C-13 and C-16 of the side chains of cercosporin are joined to form an additional six-membered ring. All the elsinochromes are optically

Elsinochrome A (**14**) R = OMe

active and show very similar CD spectra (see Fig. 4 for the spectrum of elsinochrome A (**14**)).This activity could be caused by asymmetric carbons C-1 and C-2 (and those of the side chains, if any), by a fixed noncoplanarity of the chromophore, as in cercosporin and phleichrome, or by a combination of both factors. In contrast to the ready thermal isomerization of these two compounds, elsinochrome A is quite thermostable, its optical activity remaining unchanged when a solution in chloroform is refluxed overnight (*80*).

A *cis*-orientation of the two C-acetyl groups in elsinochrome A (**14**) can be excluded. In a rigid alicyclic ring one of the benzylic protons would always be equatorial, the other axial, and hence two signals would be expected in the ^{1}H NMR spectrum, whereas the two protons give only a sharp singlet. In a rapidly interconverting system the two conformers would be enantiomorphs and the compound should not be optically active. With a rigid *trans*-configuration of the side chains, the equatorial orientation of the acetyl groups would give rise to a strong steric interaction with the methoxy groups at C-3 and C-11 which is relieved in the opposite conformer. In this latter, the two benzylic protons would be equatorial, with a dihedral angle of ca. 90°. Reexamination at 80 and 300 MHz (*5*) of the ^{13}C satellite absorption in the ^{1}H NMR spectrum of elsinochrome A shows that the coupling between H-1 and H-2 is 1.6 Hz, which is fairly consistent only with the hypothesis of a rigid *trans*-diaxial orientation of the side chains. The same value is obtained from the spectra of elsinochromes B_1, B_2, C_1 and C_2. In this situation the perylenequinone system is not planar, and some contribution of the dissymmetric chromophore to the CD absorption is expected, together with that of the asymmetric carbons. Inspection of the CD curves of the elsinochromes (Fig. 4) shows close similarity in the shape of the curves and positions of most peaks and throughs with that of cercosporin, but with large differences in intensity, especially around ca. 250–300 nm. In favor of the nonplanarity of the chromophore is also the similarity between the curves of the elsinochromes and of hypocrellin (Fig. 4) which has been shown to be nonplanar, at least in the solid state, by X-ray analysis (*135*) (Fig. 4).

The other elsinochromes (except elsinochrome D) give elsino- chrome A by oxidation. As the pairs B_1 and B_2, and C_1 and C_2, respectively, have the same structure, they must differ in the configura- tion of the CHOH for the former and of both CHOH groups in the latter. Extensive analysis of the NMR spectra of the elsinochromes corroborated by NOE experiments can give information on the relative stereochemistry of these compounds (*5*). On the other hand, the *trans*- orientation of the side chains requires the same configuration for C-1 and C-2.

Similar reasoning holds for hypocrellin, where the relative configu- ration of the asymmetric carbons has been established by X-ray analysis which also shows the existence of a definite distortion of the perylene- quinone chromophore. Again the possible CD correlation (Fig. 4) with cercosporin can give the absolute configuration.

Both the aphid pigments erythroaphin-*fb* (**4**) and erythroaphin-*sl* (**5**) contain a dihydroxyperylenequinone system fused onto two rather mobile pyran rings where no significant strain or strong non-bonding interaction seems to be present. It is therefore likely that they are planar compounds, and that the complex CD absorption (*5*) is due only to the contribution of the several asymmetric carbons of the satu- rated rings.

The existence of several asymmetric centers in the alicyclic part of the molecules can give rise to many stereoisomers. The origin, how- ever, of these compounds from two similar precursors (see Section 6.) with the same chirality reduces the number of natural isomers all of which have the same absolute configuration at the C-CH$_3$ centers. Different modes of coupling and epimerization at the benzylic centers increase the number of known erythroaphins to three (*43*). The absolute configuration of the C-CH$_3$ centers has been established by degradation (*40*) to R,R-(+)-dilactic acid (**50**) of the precursors, the protoaphins, (Scheme 4, Section 3.9.) whereas correlation with the third center has been accomplished *via* NMR spectroscopy (*42*).

In rhodoaphin-*be* (**6**) the configuration of the carbons bearing the OH groups was established by correlation with the known compounds erythroaphin-*tt* and dihydroxyerythroaphin-*fb* and by spectroscopic data (*18*).

The preferred boat-like conformation of the partially saturated ring in erythroaphin-*sl* (**5**) has been established on the basis of the values of the coupling constants and of the chemical shift of the protons of this ring, whereas a less strained pseudo-chair form was suggested for erythroaphin-*fb* (*42*).

5. Tautomerism

The parent compound of most natural perylenequinones, i.e. 4,9-dihydroxy-3,10-perylenequinone (**3**) and its derivatives can show phenol-quinone tautomerism. This process can be represented by a fast equilibrium between the four tautomers A, B, C, and D (Schema 8). Depending on the substitution on the ring, some of them may be identical.

If a fast equilibrium occurs in the substituted derivatives, the percentage of the different tautomers can only be estimated from physical measurements of parameters which are averages of the contribution of the individual tautomers.

The simplest model for these compounds is naphthazarin (**63**) (*105*), which has recently (*56, 120*) been shown to be best represented by an equilibrium where the tautomers I and II are the most stable ones, the two 1,5-quinonoid forms possessing a calculated additional energy of 25 kcal/mole (*56*).

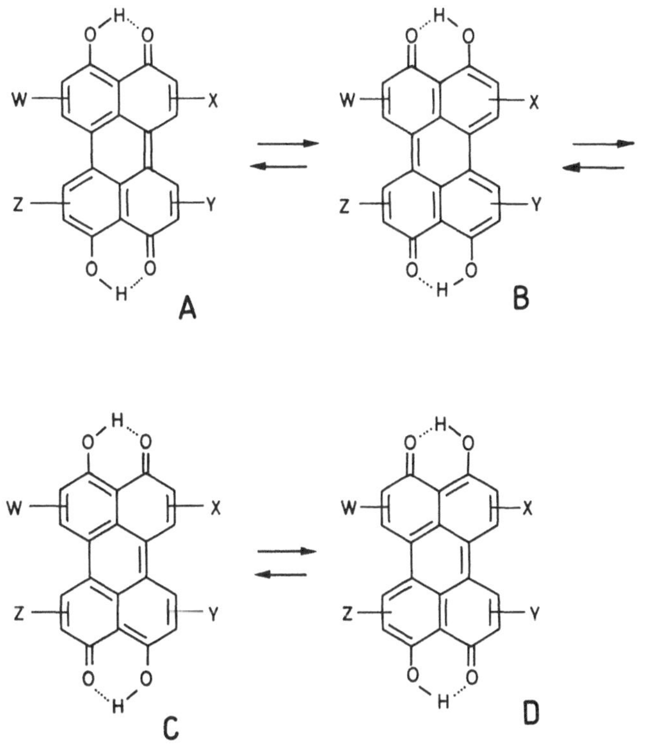

Scheme 8. Possible tautomeric forms of 4,9-dihydroxyperylene-3,10-quinone

I   (63)   II

The parent compound 4,9-dihydroxy-3,10-perylenequinone (3) does not seem to have been studied in detail in this respect. In the case of erythroaphin-*fb* (4) and rhodoaphin-*be* (6) the presence of a C_2 axis of symmetry perpendicular to the main plane of the molecule makes the two forms A and B identical, whereas in erythroaphin-*sl* (5) the different configurations of the benzylic carbons in the nonaromatic moiety require that all four forms be different. In erythroaphin-*fb* (4) the presence of only one singlet for the H-5 and H-11 protons in the ^{1}H NMR spectrum (*34*) indicates a fast equilibrium of A with B (which are identical but possess two different aromatic protons) or of C with D (which should show only one aromatic proton signal each, but are different) or of all four tautomers. The value of the chemical shift (6.61 in CDCl$_3$), which is intermediate between those for the "fixed" dimethyl ether (64) (6.3 and 6.9) and different from that for the other "fixed" dimethyl ether (65) (6.85), seems to indicate similar amounts of at least two tautomers. However, the UV spectrum of erythroaphin-*fb* (4) is so similar to that of the orange (64), while it lacks the enormously intense bands above 600 nm of the purple (65), that the predominance of tautomer A in the hydroxy compound was suggested for these solutions (*34*).

A very similar situation can be postulated for rhodoaphin-*be*, which has the same symmetry as erythroaphin-*fb*, and has ^{1}H NMR and UV spectra (*18*) very similar to those of the latter. In erythroaphin-*sl* (5) the two protons on the chromophore are nonidentical on account of the difference in the configurations of the nonchromophoric parts of the molecule; hence two different signals, 1 H each, would be expected from each single tautomer. In fact, these two signals are observed (*39*), but their chemical shifts are so similar (δ 6.58 and 6.65) and so close to that of the singlet of erythroaphin-*fb* (δ 6.61) that the tautomeric situation in both diastereoisomers must be quite analogous.

More recent and extended work has been devoted to the study of the tautomerism of the fungal perylenequinones, cercosporin and analogues, with ^{1}H and ^{13}C NMR spectrometry. In most of these com-

Erythroaphin-*fb* (**4**) R = R' = αH
Erythroaphin-*sl* (**5**) R = βH; R' = αH

(**64**)   (**65**)

pounds, tautomers C and D are identical on account of the presence of a C_2 axis of symmetry lying in the main plane of the molecule.

X-Ray data are available only for cercosporin (**18**) (*102, 108*) and hypocrellin (**41**) (*135*). They show that the former is in the tautomeric form A in the solid state, whereas the latter has a structure of type B. Obviously the situation may be different in solution, and is dependent on many variables. A systematic study of the effect of solvents, dilution and temperature (*5*) indicates that the chemical shift of significant protons (e.g. H-5 and H-8 or those of the OMe groups) may undergo large variations upon dilution. This is particularly the case for isophleichrome (**30**) and isocercosporin (**62**), where H-5 is shifted upfield as much as 1.26 and 0.37 δ, respectively, upon dilution from 10^{-1} M to 10^{-5} M solutions in chloroform. This effect might indicate for cercosporin a complete change from a prevailing tautomer B to a prevailing tautomer A; it can, however, be explained also by formation of solute-solute aggregates (5). Parameters were therefore sought which are independent of association and little sensitive to substituent effects. These are the coupling constants between the hydrogen atoms of the chelated hydroxyls and the adjacent carbons. The analysis of the spectrum of naphthazarin has given values of coupling constants ($^2J_{C8,OH} = {}^2J_{C1,OH} = 3.2$ Hz; $^3J_{C7,OH} = {}^3J_{C2,OH} = 3.8$ Hz; $^3J_{C9,OH} = 5.0$ Hz) which are the average of those of each tautomer, the process being fast on the NMR time scale. It may be safely assumed, however, that $J_{C2,OH}$ and $J_{C7,OH}$ are zero in tautomers I and II of naphthazarin, respectively, and that ^4J's with a *gauche-gauche* geometry are zero.

In substituted hydroxyperylenequinones, where there is less symmetry than in naphthazarin, all the values of the coupling constants could be obtained from analysis of ^{13}C NMR spectra (5).

From these and other data (such as chemical shifts of C-3 and C-4, and coupling constants of model compounds with "fixed" chelated carbonyls) the percentage of tautomer A and B at a certain temperature and concentration can be calculated (5).

It appears that, of the compounds studied, only isophleichrome (30) is sensitive to concentration and solvent change, whereas in cercosporin and isocercosporin tautomer A prevails completely and a ratio of ca. 1:1 holds for phleichrome and the elsinochromes. It is not yet clear which of the various contrasting factors – electronic substituent effects, steric effects (including distortion from planarity), strength of hydrogen bonds, structure and conformation of the side chains, interaction with the solvent, etc., – play major roles in stabilizing the different tautomeric forms. Some of these factors will be discussed in a forthcoming paper (5).

6. Biosynthesis

From the usual structural criteria (distribution of oxygens, etc.) all known perylenequinones obtainable from natural sources are polyketides.

The symmetrical structure of almost all natural hydroxyperylenequinones and the chemical analogy with synthetic processes suggested early (130) that the biosynthesis of these compounds occurs through phenol coupling of two polyhydroxynaphthalenes. The isolation of 4,4′,5,5′-tetrahydroxydinaphthyl (11) from the sporophores of the fungus *Daldinia concentrica*, and its enzymatic oxidation to 4,9-dihydroxy-3,10-perylenequinone (3) have been reported (1). Presumably, the dinaphthyl itself is biosynthesized by symmetrical oxidative coupling of two molecules of 1,8-dihydroxynaphthalene (**66**). This coupling reaction can be carried out *in vitro* with neutral ferricyanide (*2*), and the naphthalene itself has subsequently been isolated from *Daldinia* (*62*). In turn, the biosynthesis of 1,8-dihydroxynaphthalene via the polyketide pathway is easily visualized, so that the formation of the several metabolites may be presumed to proceed by the stages shown in Scheme 9. This interpretation is supported by the fact (*2*) that a mutant of *Daldinia* which lacks the black pigment produces the mono- (**67**) and dimethyl ethers (**68**) of 1,8-dihydroxynaphthalene, while another one, evidently blocked in the cyclization reaction which yields one of the benzene rings of the naphthalene system, excretes small amounts of 5-hydroxy-2-methylchromone (**69**) together with larger amounts of the corresponding chromanone (**70**) and with several other phenolic compounds (see Scheme 9).

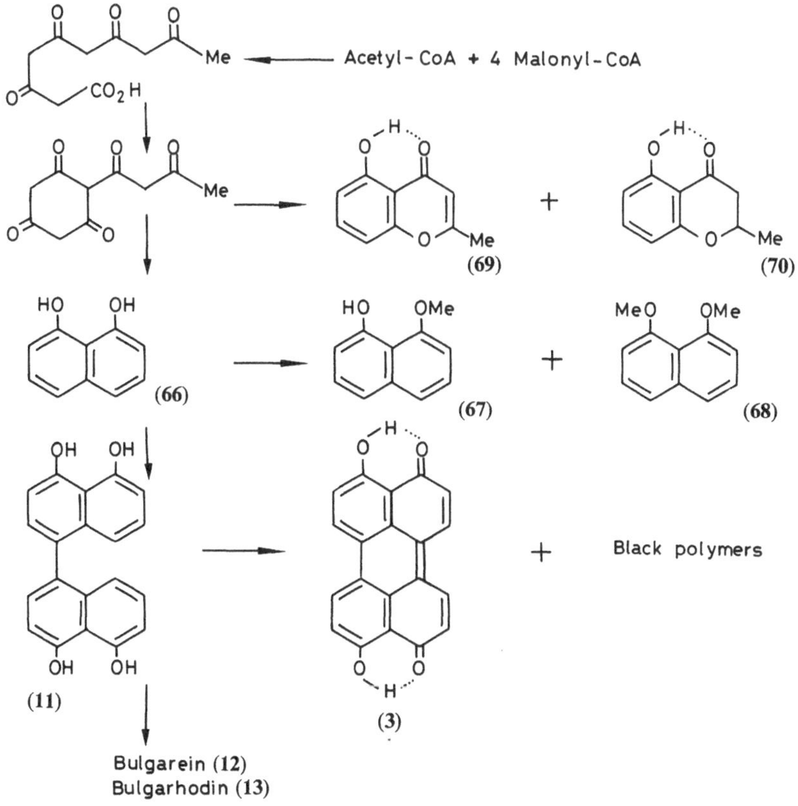

Scheme 9. Biosynthesis of aromatic metabolites in *Daldinia*

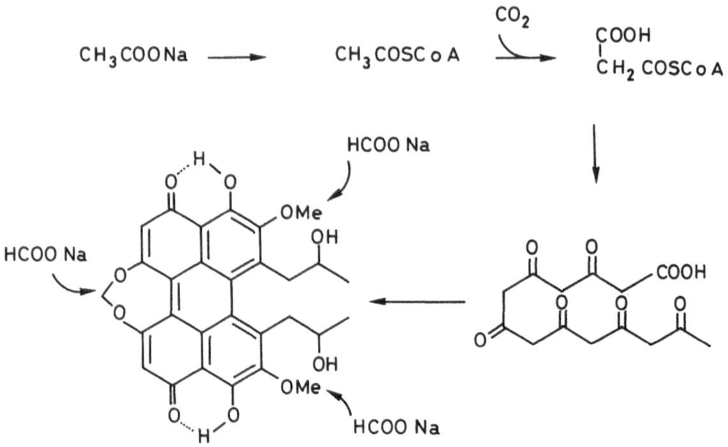

(18)

Scheme 10. Biosynthesis of cercosporin

In view of the probable biosynthesis of 4,9-dihydroxyperylene-3,10-quinone from 4,4',5,5'-tetrahydroxy-1,1'-dinaphthyl by intramolecular phenol oxidation, it is of interest that the structures of bulgarein (12) and bulgarhodin (13) (Section 3.1.) suggest their formation *in vivo* through similar intramolecular oxidation of penta- or hexahydroxy-1,1'-dinaphthyl precursors, respectively (59).

Experimental evidence for this origin has been obtained only for few of the fungal perylenequinones. Biosynthetic enrichment of elsinochrome A (14) with 1-^{14}C acetate and partial chemical degradation gave results which suggested such derivation, whereas ^{14}C formate was incorporated only into the methoxy groups (47). Similar results were obtained for cercosporin (140) and were further confirmed by successful incorporation into this compound (18) of 1-^{13}C acetate, 2-^{13}C acetate, and ^{13}C formate (109) (see Scheme 10).

However, none of these results completely excluded the possibility that elsinochrome A and cercosporin could be derived from a single polyketide chain of type B or C (Scheme 11).

The biosynthesis of elsinochromes C and D in *Pyrenochaeta terrestris* was therefore reinvestigated (76), using doubly ^{13}C-labelled acetate and 2-^{13}C, 2-[2]H$_3$-acetate, and determining the distribution of isotopes by ^{13}C and ^{2}H NMR spectrometry of elsinochrome C (16) and the acetate of elsinochrome D (17). Because of its nonsymmetrical functionality, the latter compound proved particularly suitable for this analysis. The pattern of ^{13}C labelling observed in these two pigments is

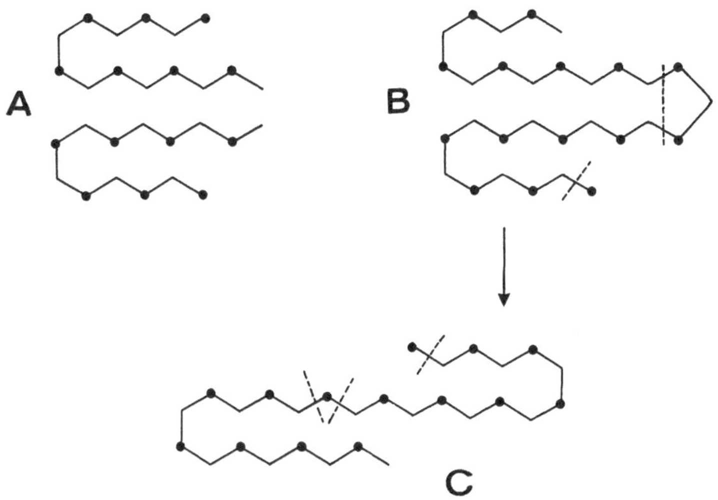

Scheme 11. Possible assembly patterns of mold perylenequinones. (Redrawn from ref. 76)

consistent with biosynthesis through formation of a naphthalenoid hep-
taketide intermediate with loss of the terminal carboxyl followed by
oxidative phenol coupling to give the complete carbon skeleton. The
findings exclude pathway B (76). Deuterium labelling was restricted
to the C-15 and C-18 methyls, a result consistent with these groups
originating in the "starter" C_2 units of the two polyketide chains and
thus excluding an initial single-chain intermediate of type C.

In one of the condensation reactions leading to the elsinochromes,
the α-carbon atoms of the two side chains R-CH$_2$-CO-Me of a hypothetical intermediate appear to be joined by some dehydrogenative process. No experimental evidence for such a reaction is yet available,
but the presumed intermediate would be the diketone corresponding
to phleichrome or isophleichrome (assuming that this condensation
is a late (76) biosynthetic step). The dimethyl ether of the analogous
diketone of the cercosporin series is known (108) and it may be significant that intramolecular aldolization of the two side chains in the postulated intermediate would account nicely for the structure of hypocrellin
(41) (Scheme 12).

All mold metabolites of group B (see Introduction) contain additional oxygenated functions on carbons 2 and 11 of the perylenequinone (3 and 12 of the elsinochromes) in addition to those on CO-derived
carbons of the initial polyketide chains. These functions must replace
carboxyls originally present (see Scheme 10). Such a replacement is
not unprecedented, but it is not common.

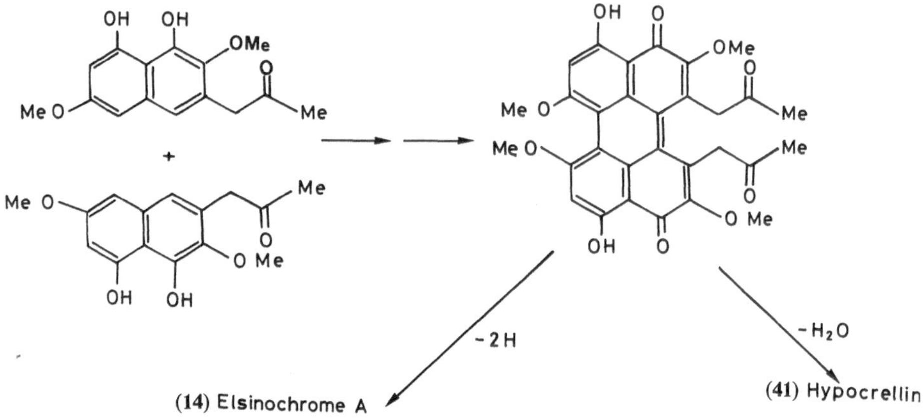

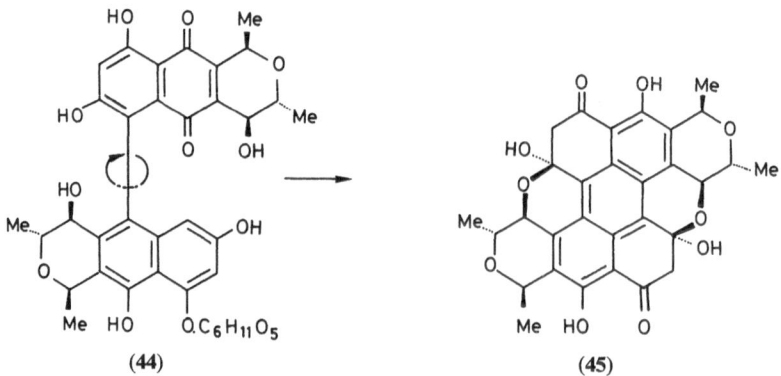

(14) Elsinochrome A   (41) Hypocrellin

Scheme 12. Possible biosynthesis of elsinochrome A and hypocrellin from a common precursor

If the two naphthyl moieties which are coupled together to form a perylenequinone system contain asymmetric carbon atoms, the question of the possible stereospecificity of the coupling, and therefore of the biosynthesis, arises. This has already been pointed out in the case of the conversion of the protoaphins to xanthoaphins, where the exclusive formation of xanthoaphin-*fb* (**45**) from protoaphin-*fb* (**44**) is supposed to occur in the one conformationally favorable sense to give the hemiketal groupings, due to hindered rotation in the noncoplanar starting material (**44**) (see Scheme 13).

(44)   (45)

Scheme 13. Cyclization of protoaphin-*fb* to xanthoaphin-*fb*

This hypothesis has been used to infer the most likely configuration of the hindered (44) (29). The cyclization step seems to be enzymatically controlled, as it proceeds under extremely mild conditions in the presence of extracts of the aphid *Dactynotus jaceae*, a species that does not produce aphin pigments (20). However, extracts of *Aphis fabae* convert protoaphin-*sl* (55) directly into xanthoaphin-*sl*-1 (56) and not *sl*-2 (57), a finding which suggests that in some cases formation of the hemiketal farthest from the binaphthyl linkage may not be enzymatically controlled (12) (Scheme 6, Section 3.9.).

A similar situation exists for fungal perylenequinones where, however, no monomer or binaphthyl intermediate other than (11) has been isolated so far. Specificity is shown by the exclusive formation *in vivo* of only one of the two diastereoisomers that are interconvertible by helix inversion, i.e. cercosporin, phleichrome, *ent*-isophleichrome, and the cladochromes. In these cases, diastereoisomeric transition states for the coupling may be envisaged, with minimum steric interaction between the side chains of the monomeric units. It may be reasonably assumed that steric interactions in a helical transition state for the coupling may be of the same type as those that occur in the final products. It appears from molecular models of cercosporin, and it is also supported by preliminary MM2 calculations on model compounds (5), that the conformation with minimum steric interactions of the side chains is the one where H-14 (or H-17), i.e. the C−H of the CHOH group, points toward the methoxyl in position 2 (or 11). Assuming that the two incoming naphthalenoid intermediates possess side chains with such a "cercosporin-like" (or "OH-exo", see Section 4.) conformation, the only favorable approach of two moieties with *R* configuration at C-14 (and C-17) gives rise to a helix with axial chirality *R*. Conversely, if the two CHOH groups have *S* configuration, the opposite helix results.

Coupling to a compound with helix *S* of two moieties with *R* configuration in the CH_2-CHOH-Me side chains occurs only if it is accompanied by a change of the conformation of the side chains to an "isocercosporin-like" one, in order to place again H-14 (and H-17) into a favorable position toward OMe-2 and thus to reduce the steric interactions to a minimum. The change is identical with that described in Section 4. for the cercosporin-isocercosporin isomerization. This is so for *ent*-isophleichrome and the cladochromes. This hypothesis is consistent with a recent elegant model study by WYNBERG (63), who has found complete stereospecificity and stereoselectivity in the coupling of two naphthol moieties with chiral side chains.

A reasonable explanation, based on the choice of the less hindered transition state, as a consequence of all three factors − nonplanar ap-

proach of the two moieties, configuration at C-14, and preferred conformation of the side chains – can thus be offered for the specificity
of the *in vivo* formation of single isomers (5).

Like the mold pigments of groups A and B, the aphins of group
C are evidently polyketides; indeed, the distribution of their oxygens
conforms completely to the one expected for acetate derived structures
(it will be remembered that this is not the case for the perylenequinones
of group B with their additional oxygens in positions 2 and 11).

The post-mortem formation of erythroaphins *fb* (4) and *sl* (5) from
their native precursors is discussed in Section 3.9. It has been elucidated
by Lord Todd and his associates. However, the ultimate biosynthetic
origin of the protoaphins, and hence that of the pigments derived from
them, is still puzzling. Their evident derivation from a polyketide makes
it extremely improbable that they should be metabolic products of
the insects themselves. Ability to form aromatic compounds *via* the
polyketide pathway has never been observed in any metazoan organism
and all present knowledge suggests that it is entirely restricted to
microbes and higher plants (*136*). The alternative possibility that these
compounds are transformation products of aromatic materials obtained from the juices of the host plants – the only food of aphids
– is similarly improbable. The protoaphins, and hence all pigments
derived from them, are made up of two linear dimethylnaphthopyran
units. While a few compounds of this type have been discovered in
some higher plants (examples: several secondary metabolites of plants
of the families Rhamnaceae, e.g. ventilagone (**71**), and Iridaceae, e.g.
isoeleutherin (**72**)) and also in a few molds, such occurrences are much
too rare and sporadic to explain the existence of aphins in a large
number of aphid species living on a wide variety of host plants. A
systematic study of one such host plant for any substance identifiable
as a likely precursor of aphins has given entirely negative results (*16*).
By exclusion, therefore, it must be assumed that the aphins (or perhaps
some aromatic precursors) are produced by the microsymbionts generally present in aphids. Experimental support for this assumption seems
to be available (*10*). It has been noted, though, that aphins are *not*
detectable in the mycetomes, the organs of the insects which serve
as shelter for the symbionts.

The only genuinely natural aphid perylenequinone is rhodoaphin-*be*
(**6**), a dihydroxyerythroaphin occurring as glycoside in a few species.
Nothing, unfortunately, seems to be known about its biosynthesis;
the irregular occurrence and scarcity of the aphids producing it has
prevented detailed study.

Strictly on structural grounds, rhodoaphin-*be* could arise plausibly
by phenol coupling of two molecules of the aglycone of neriaphin,

(71)   (72)

the ketone corresponding to glycoside B (**48**) obtained by reductive cleavage of protoaphin-*fb* (*40*) (see Section 3.9.). Neriaphin is the abundant main pigment of the orange-yellow *Aphis nerii* (*27*).

7. Biological Activity

The production of many perylenequinones by phytopathogenic fungi has stimulated hypotheses about their role in the damage done to the host plants. Their classification as toxins has been hampered by difficulties in reproducing with the pure compounds the symptoms of the disease caused by the infectious microorganism *in vivo*. The best-studied compound in this respect is cercosporin because of its wide distribution among *Cercospora* species (*7, 9, 60, 87, 106, 133*) and the economic importance of the crops attacked by this genus (beans, sugar beet, cocoa, rice, etc.).

The nonspecific toxic effect of cercosporin has suggested photodynamic activity (*79, 141*), on the basis of the similarity of its structure to that of other photodynamic agents such as hypericin, fagopyrin, and the erythroaphins. Similar activity had been observed earlier with elsinochrome A (*137*) and is also shown by phleichrome (*104, 121, 142*), isocercosporin (*90*), isophleichrome (*90*), and hypocrellin (*135*). The toxic effect of elsinochrome on bacteria and the one of cercosporin (*107, 141*), phleichrome (*121*) and hypocrellin on mice, bacteria and plant cells occurs only in the light; the action spectrum of the killing response agrees with the absorption spectrum of cercosporin (*51*), and oxygen is involved in the reaction (*51, 52, 91, 141, 144*). As a matter of fact, bixin and 1,4-diazabicyclooctane, two compounds that quench singlet oxygen (*51*), and antioxidants such as 3-*tert*-butyl-4-hydroxyanisole (BHA) and 2,6-di-*tert*-butyl-*p*-cresol (BHT) or free-radical scavengers (*91*) inhibit its activity.

The toxic effect of cercosporin was therefore attributed to the production of singlet oxygen since, besides catalyzing the formation of the endoperoxide of 2,5-dimethylfuran (*141*), a property also shown by elsinochrome A (80), it induces lipid peroxidation (55) in isolated plant and animal membranes (46) and in plant protoplasts (52, 144). Cercosporin-catalyzed photooxidation of unsaturated fatty acids produced monohydroperoxides with the hydroperoxy group attached to the carbon atoms at the position of the former double bond. This fact and the cercosporin-catalyzed bleaching of crocin (143) indicate a type II photosensitization mechanism, which is a non-radical "ene" reaction involving singlet oxygen (144). The membranes of tobacco cells and protoplasts treated with cercosporin in the light showed much of the damage associated with lipid peroxidation, such as a marked increase in the fluidity of the membranes. It also raised the membrane phase transformation temperature as determined by fatty acid spin mobility (54). These effects result in membrane damage evidenced by rapid electrolyte leakage (51, 91). Ultrastructural observation of sugar beet leaves treated with cercosporin (124) established the disruption of membranes, often the first detectable event in diseases caused by plant pathogens. The actual production of the excited $^1O_2$ species has been demonstrated by direct observation of $^1O_2$-luminescence at 1270 nm sensitized by cercosporin with a very high quantum yield (57). It has been postulated that a parasite gains an "environmental advantage" by the use of such a chemical mechanism (57).

Studies on the effect of cercosporin on plant tissues in the dark have shown inhibition of K^+ uptake and H^+ extrusion in maize roots, of K^+-Mg^{++}-dependent ATPase, and of oxidative phosphorylation, with effects on plasmalemma or alteration of the functionality of the mitochondria (92). It is suggested that this action on membrane transport phenomena adds to the photodynamic effect to cause the phytotoxic activity of cercosporin. Injection of cercosporin into healthy detached mungbean leaves increases peroxidase activity and decreases polyphenoloxidase activity with resulting decrease of total phenols, ortho-dihydroxyphenols, and flavonols (8).

Experiments directed to select cercosporin-resistant mutants in tobacco and sugar beet cell cultures, although so far unsuccessful, are being pursued actively (53). Plants resistant to paraquat appear also partially resistant to cercosporin (69).

Antimicrobial and antifungal activity has been observed for the elsinochromes (137), cercosporin, isocercosporin (100), a series of cercosporin derivatives (88), neocercosporin (99), and phleichrome (121). It seems to be observed quite generally that Gram-positive bacteria are much more sensitive to photodynamic inactivation than Gram-negative ones (15). Elsinochrome A, e.g., is highly active against S. aureus, but not at all against the Gram-negative E. coli (137), and similar results have been reported for cercosporin (100), isocercosporin (100) and neocercosporin (99). It is claimed, however, that both Gram-positive and Gram-negative bacteria are equally sensitive to the photodynamic action of cercosporin (141).

In view of the increasing interest in the therapeutic use of the photosensitizing (photodynamic) action of certain drugs (psoralen, "hematoporphyrin derivative"), it is worth mentioning that the distribution of ^{14}C-labelled cercosporin in mice has been studied (141). In the dark, no accumulation in any specific organ was observed; the compound seemed rather evenly distributed throughout the body except for the central nervous system. In light, however, stronger activity was found in the liver and in the background of the eye.

Erythroaphins or, more exactly, their precursors, are contained in the hemolymph of the insects of the family Aphididae (order Hemiptera, suborder Homoptera). The large amount (up to 1% of the weight of the live insects) must be connected with a biological function which, however, does not seem to have been established yet (43).

8. Partially Reduced Perylenequinones from Molds

Several yellow to orange pigments containing a partially reduced perylene skeleton have been isolated in recent years from the closely related phytopathogenic mold genera *Alternaria* and *Stemphylium*. They form a close-knit group of highly reactive C_{20} compounds having structures which differ from that of 4,9-dihydroxyperylene-3,10-quinone only through the presence of additional hydrogen and oxygen atoms. Their role as phytotoxins appears probable. These substances are markedly unstable, changing readily into dark, amorphous products.

8.1. Altertoxin I and Dihydroalterperylenol

Altertoxin I (**73**), reported to be toxic to bacteria, HeLa cells, and mice (68), and to be mutagenic in the Ames test (119, 123) had been observed repeatedly in isolates of various species of *Alternaria* (48, 122), in infected sorghum (117) and in other vegetables and fruits (125, 126), before it became possible to propose a structure, because of difficulties encountered in attempted isolation. It has now been isolated in pure but noncrystalline form (127). The structure (**74**) (relative ste-

(73)   (74)

reochemistry), initially advanced on the basis of 2D-^{1}H, J-resolved and ^{13}C polarization transfer NMR spectroscopy (127), should be revised to (73) since it has been shown (6) that the alleged *ortho* benzylic coupling constant (0.9–1.0 Hz) between H-6b and H-6 in structure (74), based on an erroneous assignment of the aromatic proton, is in reality a *para* benzylic coupling between H-8 and H-12a in structure (73) (6).

As a consequence of this revision, the structures of certain fluorescent products of mono- and didehydration of altertoxin I (for which highly improbable constitutions of keto tautomers of a phenol have been proposed (127)), will have to be reformulated.

More recently, a compound named dihydroalterperylenol with structure (73) was isolated (110) from strains of *Alternaria*. Proof for this structure and absolute configuration comes from analogy with alterperylenol (see below). Identity of the NMR data of both samples leads us to suggest that they are the same compound. An even more recent isolation of altertoxin I from *A. alternata* has been reported (123). Structural conclusions identical with those given above (6) in favor of structure (73) have been reached from NMR analysis. The optical rotation of altertoxin I has been given (123) for the first time; although measured in a different solvent, it indicates identity of configuration with that of dihydroalterperylenol.

8.2. Alterperylenol, Alteichin, and Altertoxin II

Alterperylenol, a reddish-orange pigment isolated, together with its dihydroderivative, from an unidentified species of *Alternaria* has the structure and absolute stereochemistry (12 R,12a S,12b S) shown in (75), substantiated by anomalous oxygen dispersion X-ray analysis (110) in addition to other spectral data.

(75)

Alterperylenol and dihydroalterperylenol show phytotoxic and anti-fungal activity. The expected acetate-based biosynthesis of alterpery-lenol, most probably *via* dimerization of a pentaketide intermediate, was confirmed by incorporation of 2-^{13}C-acetate (*110*).

Altertoxin I is always accompanied (*48, 68, 117, 122, 125, 126*) by a dehydrocompound, altertoxin II. Although ^{1}H and ^{13}C NMR spectra (*68*) and mass spectral data (*48*) were published, no structure was put forward for this substance until most recently (*123*). On the basis of extensive NMR analysis (*123*), altertoxin II was given a structure and relative stereochemistry which are identical with those of alter-perylenol (**75**), as had already been suggested (*6*). Even though they were measured in different solvents, the optical rotations reported for altertoxin II (*123*) and alterperylenol (*110*) are so similar that identity seems highly probable.

The same structure and relative stereochemistry (**75**), again on the basis of X-ray analysis, has been assigned to another (or the same?) compound isolated from *Alternaria eichhorniae* which was named altei-chin (*116*). The fungus grows on water hyacinth (*Eichhornia crassipes*), a serious pest of water ways in warm regions. Alteichin produces ne-crotic lesions resembling the pathogenic effects of the fungal infection on this and other plants. Widely different values of [α]$_D$ reported for alterperylenol and alteichin, respectively, prevent the assignment of absolute configuration to the latter or a decision on the identity of the two pigments. Alteichin is reported to be easily dehydrated to a dehydro derivative and eventually to the parent 4,9-dihydroxy-3,10-perylenequinone.

8.3. Stemphyltoxins I–IV

The stemphyltoxins are four closely related epoxy compounds pro-duced by a strain of *Stemphylium botryosum* var. *lactucum*. The struc-

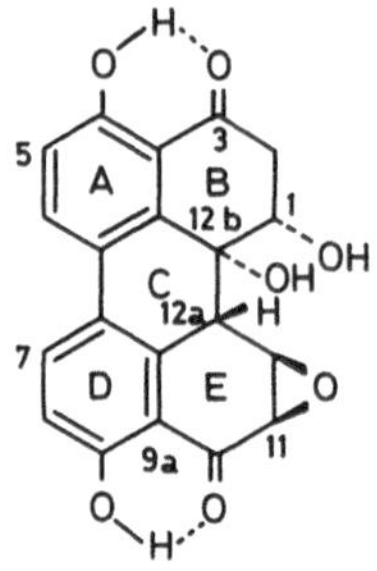

(76)

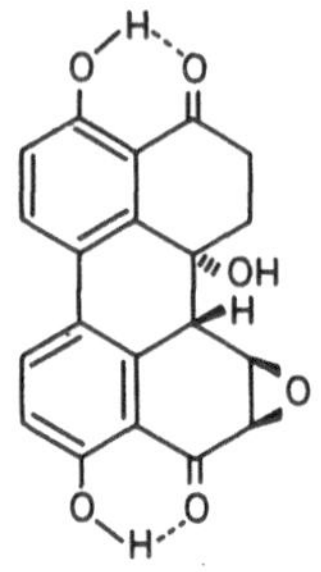

(77)

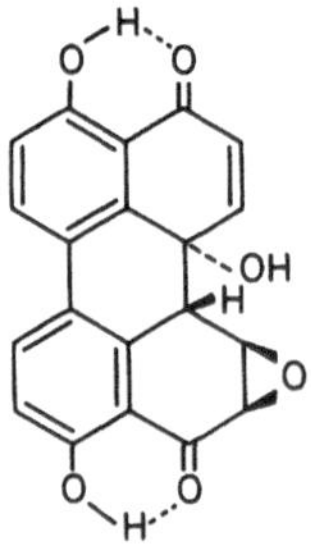

(78)

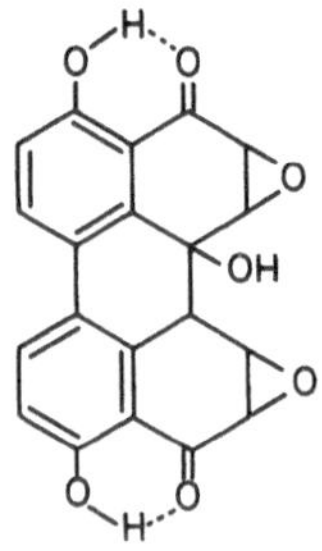

(79)

tures and relative stereochemistry $(1 S^*,11 R^*,12 R^*,12a S^*,12b S^*)$ of stemphyltoxin I (76) and II (77) were established on the basis of extensive ^1H and ^{13}C NMR studies, nuclear Overhauser effect experiments, and mass spectra (6). Comparison of the CD curves indicates that (76) and (77) also have the same absolute configuration, whereas III (78) was correlated with II (77) by catalytic hydrogenation of the double bond. The stereochemistry of IV (79), isolated in small amounts, is unknown. It is interesting to note that this group of compounds and the altertoxin-alterperylenol-alteichin group show the same relative and possibly absolute configuration at the chiral carbon atoms, thus indicating a common biosynthetic pathway. The presence of the epoxy groups may well account for the low stability and the possible phytotoxic effects of the stemphyltoxins.

All these compounds seem to be derived by a definite mode of coupling of two pentaketide units, as suggested by Okuno et al. (110) (Scheme 14), whereby a phenolic intermediate is formed and coupling occurs between two aromatic moieties, giving rise to biphenyl derivatives with the oxygenated substituents arranged in a pattern of C_{2v} symmetry. This symmetry is present in all true perylenequinones from

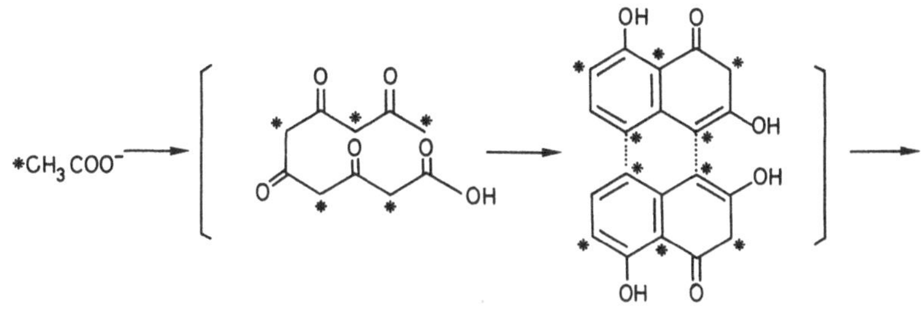

Scheme 14

molds and in most, but not in all, of the C_{20} metabolites from *Alternaria* and *Stemphylium*.

8.4. Stemphyperylenol and Altertoxin III

A mode of coupling reverse to the one just quoted in the preceding section should give rise to a dihydroanthracene derivative (cf. also the xanthoaphins and chrysoaphins). Such a structure has indeed been proposed for altertoxin I (*127*), but was disproved later (*6, 123*) (see above). It is realized, however, in stemphyperylenol, another metabolite of *Stemphylium*. Here again the structure (**80**) is supported by mass and NMR spectra. In particular, the ^{1}H NMR spectrum shows only signals for eight protons and ten carbons, indicating a symmetrical dimer. The sequences of saturated carbons are assigned from the results of decoupling experiments and the presence of NOE effects between H-6 (or 12) and H-7 (or 1). The optical activity of (**80**), combined with NMR data, requires the existence of a C_2 symmetry axis perpendicular to the main plane of the molecule, and a *cis*-relationship between H-6b and H-12b. The absolute configuration R at C-1 (and C-7) was established by application of the Horeau method of kinetic

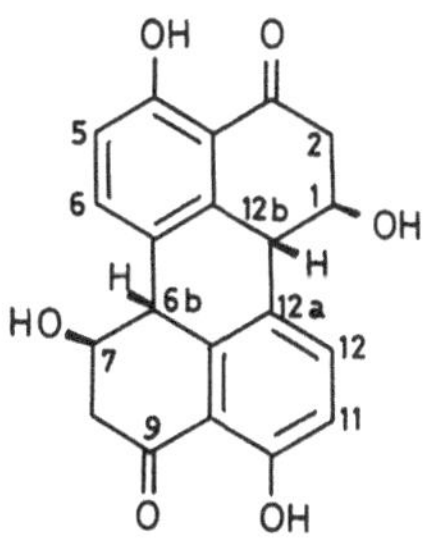

(**80**)

(81)

resolution; the relationship with the other chiral centers follows from NMR (6).

A similar line of reasoning, based on the simplicity of the NMR spectrum due to symmetry and on the optical activity, supported by $NaBH_4$ reduction to a phenolate anion less conjugated than the corresponding product from altertoxin I, has permitted elucidation of the structure and relative stereochemistry of altertoxin III (81), a metabolite of *A. alternata* (123).

Similar results have been obtained for the same compound isolated from *Stemphylium botryosum* (5). Altertoxin III is positive in the Ames test for mutagenicity (123).

Addendum

Recent X-ray analysis (5) of elsinochrome A indicates that this compound possesses a helical shape also in the solid state, such as that found in solution from NMR data.

References

1. ALLPORT, D.C., and J.D. BU'LOCK: Pigmentation and Cell-wall Material of *Daldinia* Sp. J. Chem. Soc. 4090 (1958).

2. – – Biosynthetic Pathways in *Daldinia concentrica*. J. Chem. Soc. 654 (1960).

3. ANDERSON, J.M., and J. MURRAY: Isolation of 4:9-Dihydroxy-perylene-3:10-quinone from a Fungus. Chem. and Ind. 376 (1956).

4. ARNONE, A., L. CAMARDA, G. NASINI, and L. MERLINI: Secondary Mould Metabolites. Part 13. Fungal Perylenequinones: Phleichrome, Isophleichrome, and Their Endoperoxides. J. Chem. Soc. Perkin Trans. I 1387 (1985).

5. ARNONE, A., G. ASSANTE, L. MERLINI, R. MONDELLI, G. NASINI, and U. WEISS: Unpublished work (1986).

6. ARNONE, A., G. NASINI, L. MERLINI, and G. ASSANTE: Secondary Mould Metabolites. Part 16. Stemphyltoxins, new Reduced Perylenequinone Metabolites from *Stemphylium botryosum* var. *lactucum*, J. Chem. Soc. Perkin Trans. I 525 (1986).

7. ASSANTE, G., R. LOCCI, L. CAMARDA, L. MERLINI, and G. NASINI: Screening of the Genus *Cercospora* for Secondary Metabolites. Phytochemistry **16**, 243 (1977).
8. BAJAJ, K.L., P.P. SINGH, and G. KAUR: Effect of Cercosporin Toxin on Polyphenol Metabolism in Mungbean (*Vigna radiata* L. WILCZEK) Leaves. Biochem. Physiol. Pflanzen **180**, 621 (1985).
9. BALIS, C., and M.G. PAYNE: Triglycerides and Cercosporin from *Cercospora beticola*: Fungal Growth and Cercosporin Production. Phytopathology **61**, 1477 (1971).
10. BANKS, H.J.: Ph.D. Thesis, University of Cambridge, 1969, quoted in refs. *26* and *128*.
11. BANKS, H.J., and D.W. CAMERON: Colouring Matters of the Aphididae. XXXIX. Deoxyprotoaphin. Austral. J. Chem. **25**, 2199 (1972).
12. BANKS, H.J., D.W. CAMERON, and J.C.A. CRAIK: Colouring Matters of the Aphididae. Part XXXVII. Some Further Constituents of *Tuberolachnus salignus* and their Relevance to the Stereochemistry of the Protoaphins. J. Chem. Soc. (C) 627 (1969).
13. BARBETTA, M., G. CASNATI, and A. RICCA: Aspergillina. Rend. Ist. Lombardo Sci. Lett. (A) **101**, 75 (1967).
14. BATTERHAM, T.J., and U. WEISS: The Structure of Elsinochrome A. Proc. Chem. Soc. 89 (1963).
15. BLUM, H.F.: "Photodynamic Action and Diseases Caused by Light". Reinhold Publishing Co., New York 1941.
16. BOWIE, J.H., and D.W. CAMERON: Colouring Matters of the Aphididae. Part XXV. A Comparison of Aphid Constituents with those of their Host Plants. A Glyceride of Sorbic Acid. J. Chem. Soc. 5651 (1965).
17. – – Colouring Matters of the Aphididae. Part XXVII. Mass Spectra of Aphin Derivatives. J. Chem. Soc. (B) 684 (1966).
18. – – Colouring Matters of the Aphididae. Part XXXII. Rhodoaphin. J. Chem. Soc. (C) 704 (1967).
19. – – Colouring Matters of the Aphididae. Part XXXIV. Rhodo- and Xanthodactynaphins. J. Chem. Soc. (C) 712 (1967).
20. – – Colouring Matters of the Aphididae. Part XXXV. Protodactynaphin. J. Chem. Soc. (C) 720 (1967).
21. BROCKMANN, H.: Photodynamisch Wirksame Pflanzenfarbstoffe. Fortschr. Chem. Organ. Naturstoffe **14**, 141 (1957).
22. BROWN, B.R., T. EKSTRAND, A.W. JOHNSON, S.F. MACDONALD, and A.R. TODD: Colouring Matters of the Aphididae. Part VI. The Glucosidic Nature of Protoaphin. J. Chem. Soc. 4925 (1952).
23. BROWN, B.R., A.W. JOHNSON, S.F. MACDONALD, J.R. QUAYLE, and A.R. TODD: Colouring Matters of the Aphididae. Part VII. Addition Reactions of Erythroaphin-*fb*. J. Chem. Soc. 4928 (1952).
24. BROWN, B.R., A.W. JOHNSON, J.R. QUAYLE, and A.R. TODD: Colouring Matters of the Aphididae. Part VIII. Studies on the Nature of the Aromatic Ring System in the Erythroaphins. J. Chem. Soc. 107 (1954).
25. BROWN, B.R., and A.R. TODD: Colouring Matters of the Aphididae. Part IX. Some Reactions of Extended Quinones. J. Chem. Soc. 1280 (1954).
26. BROWN, K.S., JR.: The Chemistry of Aphids and Scale Insects. Chem. Soc. Revs. **4**, 263 (1975).
27. BROWN, K.S., JR., D.W. CAMERON, and U. WEISS: Chemical Constituents of the Bright Orange Aphid, *Aphis nerii* Fonscolombe. I. Neriaphin and 6-Hydroxymusizin 8-O-β-D-Glucoside. Tetrahedron Letters 471 (1969).
28. CAHN, R.S., C. INGOLD, and V. PRELOG: Specification of Molecular Chirality. Angew. Chem. Int. Ed. **5**, 385 (1966).

29. CALDERBANK, A., D.W. CAMERON, R.I.T. CROMARTIE, Y.K. HAMIED, E. HASLAM, D.G.I. KINGSTON, LORD TODD, and J.C. WATKINS: Colouring Matters of the Aphididae. Part XX. The Structure of the Xanthoaphins and Chrysoaphins. J. Chem. Soc. 80 (1964).
30. CALDERBANK, A., A.W. JOHNSON, and A.R. TODD: Colouring Matters of the Aphididae. Part X. Preparation and Properties of 4:9-Dihydroxyperylene-3:10-quinone. J. Chem. Soc. 1285 (1954).
31. CAMERON, D.W., and H.W.-S. CHAN: Colouring Matters of the Aphididae. Part XXVIII. A Coupling Reaction involving Phenols and Quinones. Reconstitution of the Protoaphins, and Synthesis of the Chrysoaphin Chromophore. J. Chem. Soc. (C) 1825 (1966).
32. CAMERON, D.W., H.W.-S. CHAN, and E.M. HILDYARD: Colouring Matters of the Aphididae. Part XXIX. Partial Synthesis of Erythroaphin and Related Systems by Oxidative Coupling. J. Chem. Soc. (C) 1832 (1966).
33. CAMERON, D.W., H.W.S. CHAN, and D.G.I. KINGSTON: Colouring Matters of the Aphididae. Part XXIV. The Enzymic Conversion of Protoaphin into Xanthoaphin. J. Chem. Soc. 4363 (1965).
34. CAMERON, D.W., H.W.-S. CHAN, and M.R. THOSEBY: Colouring Matters of the Aphididae. Part XXXVIII. Methylation of Erythroaphin. Cationic Species Derived from Aphins in Acidic Media. J. Chem. Soc. (C) 631 (1969).
35. CAMERON, D.W., and J.C.A. CRAIK: Colouring Matters of the Aphididae. Part XXXVI. The Configuration of the Glucoside Linkage in Protoaphins. J. Chem. Soc. (C) 3068 (1966).
36. CAMERON, D.W., R.I.T. CROMARTIE, Y.K. HAMIED, E. HASLAM, D.G.I. KINGSTON, LORD TODD, and J.C. WATKINS: Colouring Matters of the Aphididae. Part XXVI. The Chrysoaphins-sl and their Reaction with Periodate. J. Chem. Soc. 6923 (1965).
37. CAMERON, D.W., R.I.T. CROMARTIE, Y.K. HAMIED, B.S. JOSHI, P.M. SCOTT, and LORD TODD: Colouring Matters of the Aphididae. Part XIX. Further Reactions of the Erythroaphins. J. Chem. Soc. 72 (1964).
38. CAMERON, D.W., R.I.T. CROMARTIE, Y.K. HAMIED, P.M. SCOTT, and LORD TODD: Colouring Matters of the Aphididae. Part XVIII. The Structure and Chemistry of the Erythroaphins. J. Chem. Soc. 62 (1964).
39. CAMERON, D.W., R.I.T. CROMARTIE, Y.K. HAMIED, P.M. SCOTT, N. SHEPPARD, and LORD TODD: Colouring Matters of the Aphididae. Part XXI. Nuclear Magnetic Resonance Evidence for the Structures of the Erythroaphins and their Derivatives. J. Chem. Soc. 90 (1964).
40. CAMERON, D.W., R.I.T. CROMARTIE, D.G.I. KINGSTON, and LORD TODD: Colouring Matters of the Aphididae. Part XVII. The Structure and Absolute Stereochemistry of the Protoaphins. J. Chem. Soc. 51 (1964).
41. CAMERON, D.W., R.I.T. CROMARTIE, D.G.I. KINGSTON, and G.B.V. SUBRAMANIAN: Colouring Matters of the Aphididae. Part XXIII. Synthesis of the Xanthoaphin Chromophore. J. Chem. Soc. 4565 (1964).
42. CAMERON, D.W., D.G.I. KINGSTON, N. SHEPPARD, and LORD TODD: Colouring Matters of the Aphididae. Part XXII. Nuclear Magnetic Resonance Evidence for the Structures and Conformations of the Naphthoquinone Dimethyl Ethers Derived from the Protoaphins, and of the Erythroaphins. J. Chem. Soc. 98 (1964).
43. CAMERON, D.W., and LORD TODD: In: Oxidative Coupling of Phenols (eds. W.I. Taylor and A.R. Battersby), Arnold, London 1967.
44. CAMERON, D.W., W.H. SAWYER, and V.M. TRIKOJUS: Colouring Matters of the Aphidoidea. XLII. Purification and Properties of the Cyclising Enzyme [Protoaphin Dehydratase (Cyclising)] concerned with Pigment Transformations in the Woolly Aphid Eriosoma lanigerum Hausmann (Hemiptera: Insecta). Austral. J. Biol. Sci. 30, 173 (1977).

45. CARLEY, H.E., R.D. WATSON, and D.M. HUBER: Inhibition of Pigmentation in *Aspergillus niger* by Dimethylsulfoxide. Canad. J. Botany **45**, 1451 (1967).
46. CAVALLINI, L., A. BINDOLI, F. MACRÌ, and A. VIANELLO: Lipid Peroxidation Inducted by Cercosporin as a Possible Determinant of its Toxicity. Chem.-Biol. Interact. **28**, 139 (1979).
47. CHEN, C.-T., K. NAKANISHI, and S. NATORI: Biosynthesis of Elsinochrome A, the Perylenequinone from *Elsinoë* spp. I. Chem. Pharm. Bull. **14**, 1434 (1966).
48. CHU, F.S.: Isolation of Altenuisol and Altertoxin I and II, Minor Mycotoxins Elaborated by *Alternaria*. J. Amer. Oil Chem. Soc. **58**, 1006 A (1981).
49. COOK, J.W., C.L. HEWETT, and I. HIEGER: The Isolation of a Cancer-producing Hydrocarbon from Coal Tar. Part II. Isolation of 1:2- and 4:5-Benzypyrenes, Perylene, and 1,2-Benzanthracene. J.Chem. Soc. 396 (1933).
50. DALLACKER, F., and H. LEIDIG: Darstellung des 4,6,7,9-Tetramethoxy-3,10-perylenchinons. Chem. Ber. **112**, 2672 (1979).
51. DAUB, M.E.: Cercosporin, a Photosensitizing Toxin from *Cercospora* species. Phytopathology **72**, 370 (1982).
52. – Peroxidation of Tobacco Membrane Lipids by the Photosensitizing Toxin, Cercosporin. Plant Physiol. **69**, 1361 (1982).
53. – A Cell Culture Approach for the Development of Disease Resistance: Studies on the Phytotoxin Cercosporin. HortScience **19**, 382 (1984).
54. DAUB, M.E., and S.P. BRIGGS: Changes in Tobacco Cell Membrane Composition and Structure Caused by Cercosporin. Plant Physiol. **71**, 763 (1983).
55. DAUB, M.E., and R.P. HANGARTER: Light-Induced Production of Singlet Oxygen and Superoxide by the Fungal Toxin, Cercosporin. Plant Physiol. **73**, 855 (1983).
56. DE LA VEGA, J.R., J.H. BUSCH, J.H. SCHAUBLE, K.L. KUNZE, and B.E. HAGGERT: Symmetry and Tunneling in the Intramolecular Proton Exchange in Naphthazarin, Methylnaphthazarin and Dimethylnaphthazarins. J. Amer. Chem. Soc. **104**, 3295 (1982).
57. DOBROWOLSKI, D.C., and C.S. FOOTE: Cercosporin, a Singlet Oxygen Generator. Angew. Chem. Int. Ed. Engl. **22**, 720 (1983).
58. DUEWELL, H., A.W. JOHNSON, S.F. MACDONALD, and A.R. TODD: Colouring Matters of the Aphididae. Part III. Colouring Matters of *Tuberolachnus salignus*. J. Chem. Soc. 485 (1950).
59. EDWARDS, R.L., and H.J. LOCKETT: Constituents of the Higher Fungi. Part XVI. Bulgarhodin and Bulgarein, Novel Benzofluoranthenequinones from the Fungus *Bulgaria inquinans* (Fries). J. Chem. Soc. Perkin Trans. I, 2149 (1976).
60. FAJOLA, A.O.: Cercosporin, a Phytotoxin from *Cercospora* spp. Physiol. Plant Pathol. **13**, 157 (1978).
61. FATIADI, A.J.: Preparation and Properties of some Oxidation Products of Perylene. J. Res. Nat. Bur. Stand. **72 A**, 39 (1968).
62. FATTORUSSO, E., M. PIATTELLI, and R.A. NICOLAUS: Su alcune melanine naturali. Rend. Accad. Sci. Fis. Mat. (Soc. Naz. Sci. Lett. Arti Napoli) **32**, 200 (1965).
63. FERINGA, B., and H. WYNBERG: Asymmetric Phenol Oxidation. Stereospecific and Stereoselective Oxidative Coupling of a Chiral Tetrahydronaphthol. J. Org. Chem. **46**, 2547 (1981).
64. FREDGA, A.: Steric Correlations by the Quasi-racemate Method. Tetrahedron **8**, 126 (1960).
65. HACKENG, W.H.L., H. COPIER, and C.A. SALEMINK: On the Structure of Phycaron. Rec. Trav. Chim. Pays-Bas **82**, 322 (1963).
66. HACKENG, W.H.L.: Elsinochroom A. Ph. D. Thesis, Utrecht 1963.
67. HALL, D.M.: Stereochemistry of 2,2′-Bridged Biphenyls. In: Progress in Stereochemistry (B.J. HAYLETT and W.A. HARRIS eds.), **4**, 1–42, London, Butterworths 1969

68. HARVAN, D.J., and R.W. PERO: In: Mycotoxins and Other Fungal Related Food Problems, (J.V. Rodricks ed.), Adv. Chem. Ser. No. 149, p. 344, American Chemical Society, Washington D.C., 1976.

69. HUGHES, K.W., D. NEGROTTO, M.E. DAUB, and R.L. MEEUSEN: Free-radical Stress Response in Paraquat-sensitive and Resistant Tobacco Plants. Environm. Exp. Bot. **24**, 151 (1984).

70. HUMAN, J.P.E., A.W. JOHNSON, S.F. MACDONALD, and A.R. TODD: Colouring Matters of the Aphididae. Part II. Colouring Matters from *Aphis fabae*. J. Chem. Soc. 477 (1950).

71. JENKINS, A.E., and A.A. BITANCOURT: An *Elsinoe* causing an Anthracnose on *Hickoria pecan*. Phytopathology **28**, 75 (1938).

72. – – Studies on the Myriangiales. XIV. *Phyllosticta caryae* Rand *non* Pk. as *Sphaceloma*, including its Separation from Peck's species. Arquivos do Instituto Biológico São Paulo) **32**, 61 (1965).

73. JENKINS, A.E.: Private Communication to U. Weiss.

74. JOHNSON, A.W., J.R. QUAYLE, T.S. ROBINSON, N. SHEPPARD, and A.R. TODD: Colouring Matters of the Aphididae. Part V. Infra-red Spectra. J. Chem. Soc. 2633 (1951).

75. JOHNSON, A.W., SIR ALEXANDER R. TODD, and J.C. WATKINS. Colouring Matters of the Aphididae. Part XV. The Alkaline Inversion of Erythroaphin-*sl* and its Derivatives. J. Chem. Soc. 4091 (1956).

76. KUROBANE, I., L.C. VINING, A.G. MCINNES, D.G. SMITH, and J.A. WALTER: Biosynthesis of Elsinochromes C and D. Pattern of acetate incorporation determined by ^{13}C and ^{2}H nmr. Canad. J. Chem. **59**, 422 (1981).

77. KUYAMA, S.: Cercosporin. A Pigment of *Cercosporina Kikuchii* Matsumoto et Tomoyasu. III. The Nature of the Aromatic Ring of Cercosporin. J. Org. Chem. **27**, 939 (1962).

78. KUYAMA, S., and T. TAMURA: Cercosporin. A Pigment of *Cercosporina Kikuchii* Matsumoto et Tomoyasu. I. Cultivation of Fungus, Isolation and Purification of Pigment. J. Amer. Chem. Soc. **79**, 5725 (1959).

79. – – Cercosporin. A Pigment of *Cercosporina Kikuchii* Matsumoto et Tomoyasu. II. Physical and Chemical Properties of Cercosporin and its Derivatives. J. Amer. Chem. Soc. **79**, 5726 (1957).

80. LOUSBERG, R.J.J.CH., C.A. SALEMINK, U. WEISS, and T.J. BATTERHAM: Pigments of *Elsinoe* Species. Part II. Structure of Elsinochromes A, B, and C. J. Chem. Soc. (C) 1219 (1969).

81. LOUSBERG, R.J.J.CH., U. WEISS, and C.A. SALEMINK: Pigments of *Elsinoe* Species. Part III. Methylation of Elsinochrome A. Formation of Mono- and Di-methyl Ethers. J. Chem. Soc. (C) 2152 (1970).

82. LOUSBERG, R.J.J.CH., C.A. SALEMINK, and U. WEISS: The Pigments of *Elsinoe* Species. Part V. The Structure of Elsinochrome D. J. Chem. Soc. 2159 (1970).

83. LOUSBERG, R.J.J.CH.: Pigments of *Elsinoe* Species. Ph. D. Thesis, Utrecht, 1969.

84. LOUSBERG, R.J.J.CH., L. PAOLILLO, H. KON, U. WEISS, and C.A. SALEMINK: Pigments of *Elsinoe* Species. Part IV. Confirmatory Evidence for the Structure of Elsinochrome A and its Ethers from Studies of Nuclear Magnetic Resonance (Solvent and Overhauser Effects) and Electron Spin Resonance. J. Chem. Soc. (C) 2154 (1970).

85. LOUSBERG, R.J.J.CH., U. WEISS, C.A. SALEMINK, A. ARNONE, L. MERLINI, and G. NASINI: The Structure of Cercosporin, a Naturally Occurring Quinone. Chem. Commun. 1463 (1971).

86. LUND, N.A., A. ROBERTSON, and W.B. WHALLEY: The Chemistry of Fungi. Part XXI. Asperxanthone and a Preliminary Examination of Aspergillin. J. Chem. Soc. 2434 (1953).

87. LYNCH, F.J., and M.J. GEOGHEGAN: Production of Cercosporin by *Cercospora* Species. Trans. Brit. Mycol. Soc. **69**, 496 (1977).
88. – – Antibiotic Activity of a Fungal Perylene-quinone and some of its Derivatives. Trans. Brit. Mycol. Soc. **72**, 31 (1979).
89. MACDONALD, S.F.: Colouring Matters of the Aphididae. Part XI. Pigments from *Hamamelistes* Species. J. Chem. Soc. 2378 (1954).
90. MACRÌ, F., and A. VIANELLO: Photodynamic Activity of Compounds Structurally Related to Cercosporin. Agr. Biol. Chem. **44**, 2967 (1980).
91. – – Photodynamic activity of cercosporin on plant tissues. Plant, Cell and Environm. **2**, 267 (1979).
92. – – Inhibition of K^+ Uptake and H^+ Extrusion Caused by Non-irradiated Cercosporin. Plant Sci. Letters **22**, 29 (1981).
93. MAHADEVAN, A., J. KUČ, and E.B. WILLIAMS: Biochemistry of Resistance in Cucumber against *Cladosporium cucumerinum*. I. Presence of a Pectinase Inhibitor in Resistant Plants. Phytopathology **55**, 1000 (1965).
94. MATSUEDA, S.: Private communication (1984).
95. – Influence of light on pigment formation in *Cercosporina kikuchii*. Seikagaku **41**, 714 (1978).
96. – Structure of Neosporin. Chem. and Ind. 233 (1978).
97. MATSUEDA, S., M. NAGAKI, Y. SUSUTA, K. TANAKA, K. TAKAGAKI, M. SHIMOYAMA, T. IMAIZUMI, and K. TSUBAKI: *Cercospora kikuchii* (Matsumoto et Tomoyasu) Gardner mut. *alba*, a Novel Mutant of the Pathogen of the Soybean Purple Speck Disease. Sci. Rep. Hirosaki Univ. **30**, 42 (1983).
98. MATSUEDA, S., R. TAKAHASHI, Y. MASUCHI, K. TAKAGAKI, and M. SHIMOYAMA: Studies on Fungal Products. III. Structure of Neocercosporin. Yakugaku Zasshi **98**, 1553 (1978).
99. MATSUEDA, S., Y. MASUCHI, K. TAKAGAKI, M. SHIMOYAMA, R. TAKAHASHI, and T. SATOMI: Studies on Fungal Products. IV. Antimicrobial Aspects of Neocercosporin. Yakugaku Zasshi **99**, 20 (1979).
100. MATSUEDA, S., K. TAKAGAKI, M. SHIMOYAMA, and A. SHIOTA: Studies on Fungal Products. V. Antimicrobial Aspects of Quinone Derivatives. Yakugaku Zasshi **100**, 900 (1980).
101. MATSUEDA, S., K. TAKAGAKI, M. SHIMOYAMA, T. IMAIZUMI, and M. KOREEDA: Structure of Amphicercosporin and Protocercosporin. Chem. & Ind. 58 (1982).
102. MENTZAFOS, D., A. TERZIS, and S.E. FILIPPAKIS: 1,12-Bis-(2-Hydroxypropyl)-2,11-dimethoxy-6,7-methylenedioxy-4,9-dihydroxyperylen-3,10-quinone. Ethanol. Water. Cryst. Struct. Comm. **11**, 71 (1982).
103. MERLINI, L., and G. NASINI: Natural Perylenequinones from Moulds. In: "Chemistry and Biotechnology of Biologically Active Natural Products", Cs. Szàntay ed., Akadémiai Kiadò, Budapest 1983, p. 121.
104. MINO, Y., T. IDONUMA, and R. SAKAI: Effect of Phleichrome Produced by the Timothy Leaf Spot Fungus, *Cladosporium phlei* on the Invertase from the Host Leaves. Ann. Phytopathol. Soc. Japan **45**, 463 (1979).
105. MOORE, R.E., and P.J. SCHEUER: Nuclear Magnetic Resonance Spectra of Substituted Naphthoquinones. Influence of Substituents on Tautomerism, Anisotropy and Stereochemistry in the Naphthazarin System. J. Org. Chem. **31**, 3272 (1966).
106. MUMMA, R.O., F.L. LUKEZIC, and M.G. KELLY: Cercosporin from *Cercospora hayii*. Phytochemistry **12**, 917 (1973).
107. MUTTO, S., and V. D'AMBRA: Effetti della Cercosporina su Cellule Fogliari di *Beta vulgaris* var. *saccharifera*. Riv. Patologia Vegetale [4], **17**, 71 (1981).
108. NASINI, G., L. MERLINI, G.D. ANDREETTI, G. BOCELLI, and P. SGARABOTTO: Stereochemistry of Cercosporin. Tetrahedron **38**, 2787 (1982).

109. OKUBO, A., S. YAMAZAKI, and K. FUWA: Biosynthesis of Cercosporin. Agr. Biol. Chem. **39**, 1173 (1975).
110. OKUNO, T., I. NATSUME, K. SAWAI, K. SAWAMURA, A. FURUSAKI, and T. MATSU-MOTO: Structure of Antifungal and Phytotoxic Pigments produced by *Alternaria* sps. Tetrahedron Letters **24**, 5653 (1983).
111. OVEREEM, J.C., A.K. SIJPESTEIJN, and A. FUCHS: The Formation of Perylenequinones in Etiolated Cucumber Seedlings infected with *Cladosporium cucumerinum*. Phytochemistry **6**, 99 (1967).
112. QUILICO, A.: Sulla Natura del Pigmento delle Spore di *Aspergillus niger*. Nota III sull'Aspergillina. Gazz. Chim. Ital. **63**, 400 (1933).
113. RAND, F.V.: Some Diseases of Pecans. J. Agric. Res. **1**, 303 (1914).
114. RAWAT, A.K.: Some Observations on the Aspergillin of *Aspergillus niger*. Arch. Biochem. Biophys. **124**, 418 (1968).
115. RAY, A.C., and R.E. EAKIN: Studies on the Biosynthesis of Aspergillin by *Aspergillus niger*. Applied Microbiology **30**, 909 (1975).
116. ROBESON, D., G. STROBEL, G.K. MATUSUMOTO, E.L. FISHER, M.H. CHEN, and J. CLARDY: Alteichin: an Unusual Phytotoxin from *Alternaria eichorniae,* a Fungal Pathogen of Water Hyacinth. Experientia **40**, 1248 (1984).
117. SAUER, D.B., L.M. SEITZ, R. BURROUGHS, H.E. MOHR, J.L. WEST, R.J. MILLERET, and H.D. ANTHONY: Toxicity of *Alternaria* Metabolites Found in Weathered Sorghum Grain at Harvest. J. Agric. Food Chem. **26**, 1380 (1980).
118. SCHOLL, R., CHR. SEER, and R. WEITZENBÖCK: Perylen, ein hoch kondensierter aromatischer Kohlenwasserstoff $C_{20}H_{12}$. Ber. dtsch. chem. Ges. **43**, 2202 (1910).
119. SCOTT, P.M., and D.R. STOLZ: Mutagens Produced by *Alternaria alternata*. Mutat. Res. **78**, 33 (1980).
120. SHIAU, W.-I., E.N. DUESLER, I.C. PAUL, D.Y. CURTIN, W.G. BLANN, and C.A. FYFE: Investigation of Crystalline Naphthazarin B by ^{13}C NMR Spectroscopy Using "Magic Angle" Spinning Techniques and by X-ray Diffraction: Evidence for a Dynamic Disordered Structure. J. Amer. Chem. Soc. **102**, 4546 (1980).
121. SHIMANUKI, T., and T. ARAKI: Phleichrome, a Non-Host Specific Toxin Produced by the Causal Organism Of Timothy Purple Spot, *Cladosporium phlei*, and its Toxic Influence on Timothy Leaf Blades and Leaf Surface Microorganism. J. Japan. Soc. of Grassland Sci. **28**, 426 (1983).
122. SLIFKIN, M.K., A. OTTOLENGHI, and H. BROWN: Effects of Fungicides on Mycotoxin Production by *Alternaria mali*. Mycopathol. Mycol. Appl. **50**, 241 (1973).
123. STACK, M.E., E.P. MAZZOLA, S.W. PAGE, A.E. POHLAND, R.J. HIGHET, M.S. TEMPESTA, and D.G. CORLEY: Mutagenic Perylenequinone Metabolites of *Alternaria alternata*: Altertoxins I, II, and III. J. Natural Products **49**, 866 (1986)
124. STEINKAMP, M.P., S.S. MARTIN, L.L. HOEFFERT, and E.G. RUPPEL: Ultrastructure of Lesions Produced in Leaves of *Beta vulgaris* by Cercosporin, a Toxin from *Cercospora beticola*. Phytopathology **71**, 1272 (1981).
125. STINSON, E.E., D.D. BILLS, S.F. OSMAN, J. SICILIANO, M.J. CEPONIS, and F.G. HEISLER: Mycotoxin Production by *Alternaria* Species Grown on Apples, Tomatoes and Blueberries. J. Agric. Food Chem. **28**, 960 (1980).
126. STINSON, E.E., S.F. OSMAN, E.G. HEISLER, J. SICILIANO, and D.D. BILLS: Mycotoxin Production in Whole Tomatoes, Apples, Oranges and Lemons. J. Agric. Food Chem. **29**, 790 (1981).
127. STINSON, E.E., S.F. OSMAN, and P.E. PFEFFER: Structure of Altertoxin I, a Mycotoxin from *Alternaria*. J. Org. Chem. **47**, 4110 (1982).
128. THOMSON, R.H.: Naturally Occurring Quinones, Second Edition. Academic Press, London and New York, 1971, p. 576.
129. TODD, A.R.: Die Farbstoffe der Blattläuse (Aphididae). Experientia **18**, 433 (1962).

130. – Some New Developments in the Chemistry of Natural Colouring Matters. Chem. Brit. **2**, 428 (1966).
131. TREIBS, A., R. WILHELM, and K. JACOB: Der Quincyte-Farbstoff. Liebigs Ann. Chem. 842 (1981).
132. VAN DER VIJVER, L.M., and K.W. GERRITSMA: Naphthoquinones from Ebenaceae. Pharm. Weekblad **111**, 1273 (1976).
133. VENKATARAMANI, K.: Isolation of Cercosporin from *Cercospora personata*. Phytopathol. Z. **58**, 379 (1967).
134. WATTS, C.D., B.R. SIMONEIT, J.R. MAXWELL, and J.P. RAGOT: The Quincyte Pigments: A Novel Series of Fossil "Dyes" from an Eocene Sediment. Adv. Org. Geochem., Proc. 7th Internat. Meeting 223 (1977).
135. WEI-SHIN, C., C. YUAN-TENG, W. XIANG-YI, E. FRIEDRICHS, H. PUFF, and E. BREIT-MAIER: Die Struktur der Hypocrellins und seines Photooxidationsproduktes Peroxy-hypocrellin. Liebigs Ann. Chem. 1880 (1981).
136. WEISS, U., and J.M. EDWARDS: The Biosynthesis of Aromatic Compounds. Wiley-Interscience, New York 1980.
137. WEISS, U., H. FLON, and W.C. BURGER: The Photodynamic Pigment of some Species of *Elsinoë* and *Sphaceloma*. Arch. Biochem. Biophys. **69**, 311 (1957).
138. WEISS, U., H. ZIFFER, T.J. BATTERHAM, M. BLUMER, W.H.L. HACKENG, H. COPIER, and C.A. SALEMINK: Pigments of *Elsinoë* Species; Isolation of Pure Elsinochromes A, B, and C. Canad. J. Microbiol. **11**, 57 (1965).
139. WOLFBEIS, O.S., and E. FÜRLINGER: Absorption, Fluorescence and Fluorimetric Detection Limits of Naturally Occurring Quinonoid Antibiotics and Dyes. Mikrochim. Acta 385 (1983 III).
140. YAMAZAKI, S., and T. OGAWA: The Chemistry and Stereochemistry of Cercosporin. Agr. Biol. Chem. **36**, 1707 (1972).
141. YAMAZAKI, S., A. OKUBO, Y. AKIYAMA, and K. FUWA: Cercosporin, a Novel Photodynamic Pigment Isolated from *Cercospora kikuchii*. Agr. Biol. Chem. **39**, 287 (1975).
142. YOSHIHARA, T., T. SHIMANUKI, T. ARAKI, and S. SAKAMURA: Phleichrome; a New Phytotoxic Compound Produced by *Cladosporium phlei*. Agr. Biol. Chem. **39**, 1683 (1975).
143. YOUNGMAN, R.J., and E.F. ELSTNER: Primary Photodynamic Reactions Occurring During the Breakdown of Photosynthetic Pigments. Ber. Deutsch. Bot. Ges. **96**, 357 (1983).
144. YOUNGMAN, R.J., P. SCHIEBERLE, H. SCHNABL, W. GROSCH, and E.F. ELSTNER: The photodynamic generation of singlet molecular oxygen by the fungal phytotoxin, cercosporin. Photobiochem. Photobiophys. **6**, 109 (1983).
145. ZINKE, A., W. HIRSCH, and E. BROZEK: Perylen und Derivate. XIX. Monatsh. Chem. **51**, 205 (1929).
146. ZINKE, A., and R. WENGER: Perylene and its Derivatives. Degradation of Perylene to Benzanthrone. Monatsh. Chem. **56**, 143 (1930) and preceding papers.

(*Received December 16, 1986*)

The Pigments of the Flexirubin-Type.
A Novel Class of Natural Products

By HANS ACHENBACH, Department of Pharmaceutical Chemistry,
Institute of Pharmacy and Food Chemistry, University of Erlangen,
Erlangen, Federal Republic of Germany

With 23 Figures

Contents

I. Introduction

During their investigations on the pigments of a certain strain of
Flexibacter elegans REICHENBACH et al. observed a colouring component showing chromatographic and spectroscopic behaviour significantly different from all the carotenoids hitherto known as constituents
in microorganisms of this type (*1*). The colouring compound was called
flexirubin.

74

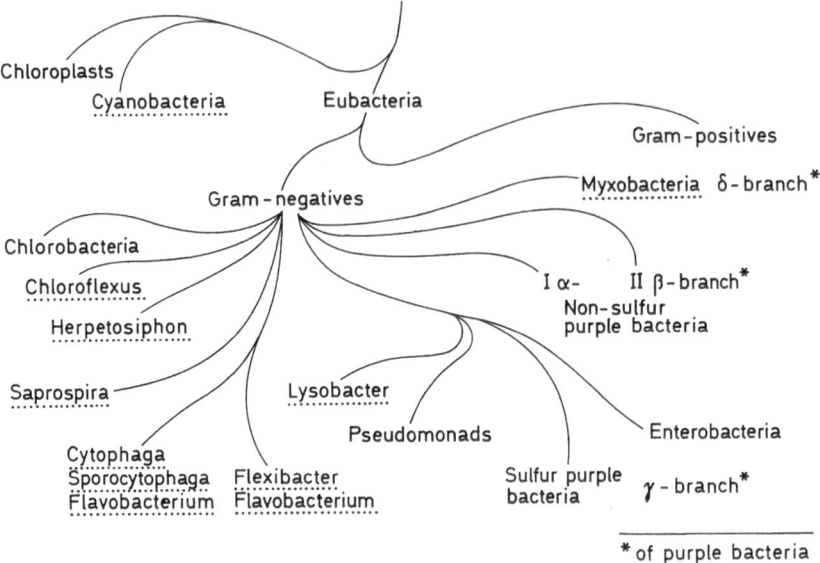

Fig. 1. Phylogenetic relationship among bacteria (gliding bacteria are underlined) (2, 3)

The *Flexibacteria* belong to the class of Gram-negative bacteria. Phyllogenetically they are related to the *Cytophaga/Sporocytophaga* group. In addition, there exists a close relationship of the *Flexibacteria* to the *Flavobacteria* (Fig. 1) (2–4).

Whereas Gram-positive bacteria generally are either non-motile or motile by means of flagellae, there are among the Gram-negative bacteria the so-called "gliding bacteria", which exhibit a gliding movement when in contact with a suitable interface in spite of the fact that they do not have flagellae. The *Flexibacteria* are gliding bacteria: they are characteristically unicellular and organotrophic. They are ubiquitous and important ecologically for the degradation of biopolymers.

Among the gliding bacteria, *Myxobacteria*, bacteria of the *Flexibacter/Cytophaga* group, and the *Flavobacteria* are known to produce pigments (5, 6). KLEINIG and REICHENBACH reported studies on the pigments of some *Myxobacteria* which they found to be fatty acid esters of the glycosidic carotenoid myxobactone (1) (6).

The only earlier detailed chemical study of pigmentation in the *Flexibacter/Cytophaga* group was carried out by AASEN and LIAAEN-JENSEN with an unknown species of *Flexibacter* (7, 8). Among other carotenoids, they isolated flexixanthin (2) and deoxyflexixanthin (3)

References, pp. 109–111

(1) $R^1 = H$ $R^2 = C_6H_{11}O_5$
(2) $R^1 = OH$ $R^2 = H$
(3) $R^1 = R^2 = H$

which differ from myxobactone mainly by their non-glycosidic character.

Around 1970, Lewin and Lounsbery (9) and Simon and White (10) studied the pigments of *Flexibacter elegans*. Obviously without isolation of chemically pure substances both groups came to the conclusion that the pigments of *Fx. elegans* are also carotenoids. It seems likely that this error was caused mainly by reliance on electron-spectroscopic data since there is indeed a close similarity between the electronic spectra of the pigments of *Fx. elegans* and the carotenoids.

In 1974 the main pigment from *Fx. elegans* was isolated by Reichenbach *et al.* and recognized as having a new, non-carotenoid structure (1). The pigment was therefore named flexirubin.

In the early stages of the investigation mass spectrometric studies contributed substantially to the conclusion that the pigment was not a carotenoid. On electron bombardment flexirubin very characteristically eliminates benzene from the molecular ion, whereas mass spectrometric fragmentation of carotenoids typically results in loss of toluene and xylene (11).

Further studies demonstrated the presence of flexirubin-type pigments in some *Cytophaga* species – mostly as hard-to-separate mixtures of up to 25 individual compounds which are accompanied by minor amounts of carotenoids (12–14).

Very recently flexirubin-type pigments have also been found in *Sporocytophaga myxoides* (15) and in *Flavobacterium* (16, 17), a group of Gram-negative bacteria, whose taxonomy is under discussion.

II. Structure Elucidation of Flexirubin

Whereas currently the physico-chemical properties of flexirubin-type pigments and in particular the mass and nmr spectrometric behaviour provide very powerful tools for rapid structure determination of even small amounts of somewhat impure samples of this novel class of natural pigments, the original structure elucidation of flexirubin was based on chemical degradation together with spectroscopic investigations (*18*). The most important steps of the chemical conversions which were performed with about 30 mg of flexirubin are outlined in Fig. 2.

High resolution mass spectrometry yielded the elementary composition $C_{43}H_{54}O_4$. The electronic spectrum corresponds to that of an ω-phenyloctaenecarboxylic acid (*19*). A reversible bathochromic shift in alkali medium indicated presence of phenolic OH-group(s). Accordingly, flexirubin was treated with methylating agents and was thus converted into a dimethyl derivative (UV/VIS: λ_{max} 449 nm, no shift with alkali; ^{1}H-nmr: new signals at δ_{OCH_3} 3.85 and 3.82). Whereas use of diazomethane formed a number of by-products, the methylation occurred quantitatively with methyl iodide in acetone. Because of its lower polarity and better solubility in organic solvents, the dimethyl derivative (**4**) is more suitable for mass spectrometric or nuclear magnetic resonance studies.

The presence of the conjugated octaene moiety – already indicated by the electronic spectra – was corroborated by catalytic hydrogenation of (**4**) with palladium/charcoal, which resulted in a colourless compound by uptake of 8 mol of hydrogen. The ^{1}H-nmr spectra of flexirubin and its dimethyl derivative exhibit only two singlets characteristic of methyl groups linked to double bonds (δ_{CH_3} at 2.3 and 2.2). Since these signals remain essentially stationary after hydrogenation, they must belong to methyl substituents on aromatic systems and, therefore, allylic methyl groups – typically present in carotenoids – could be excluded.

The following additional structural information was deduced from changes in the ^{1}H-nmr spectrum upon hydrogenation:

a) Resonances of 16 olefinic protons shifted from the range δ 5–7 to high field leaving in the aromatic region signals of only five aromatic protons, which according to their coupling patterns could be attributed to a 1,3,4- and a 1,2,4,5- or 1,2,3,5-substituted benzene ring.

b) From the original olefinic protons there appeared inter alia an (additional) multiplet (4H/around δ 2.5), which could be assigned to benzylic methylene groups or methylenes α to a carboxyl group.

These observations were in agreement with conversion of the ω-phenylheptadecaoctaenoic acid unit into the corresponding ω-phenylheptadecanoic acid.

A band at $\nu_{c=0} = 1715$ cm^{-1} in the infrared spectra of flexirubin and flexirubin dimethyl ether could be attributed to an ester group in conjugation with (an) olefinic double bond(s). This band was shifted to $\nu_{c=0} = 1750$ cm^{-1} in the hydrogenation products.

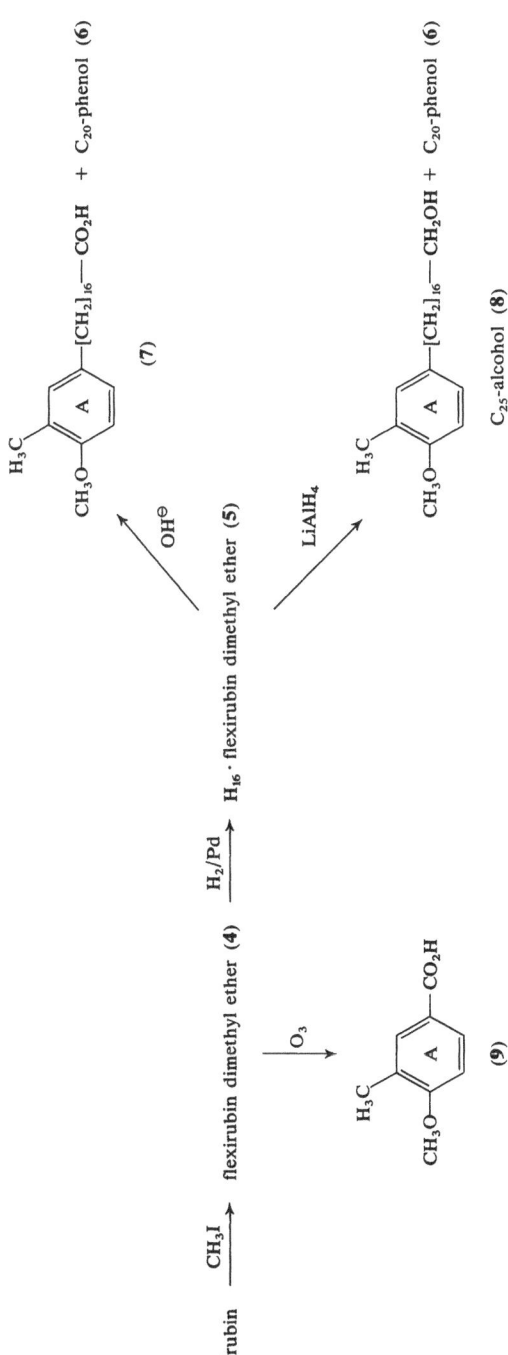

Fig. 2. Degradation of flexirubin

Cleavage of the ester produced two major products: alkaline hydrolysis of flexirubin dimethyl ether (4) resulted in an almost insoluble coloured acid and a C_{20}-phenol (6); the same treatment applied to the hydrogenated dimethyl ether (5) gave in significantly better yields a readily soluble C_{25}-ω-phenyl fatty acid (7) as well as (6).

It should be pointed out, that cleavage at the ester bond also is a major fragmentation path in the mass spectra of flexirubin, its dimethyl derivative or their hydrogenation products. In all these compounds the C_{20}-phenol as well as the corresponding C_{25}-acid represent key fragments (Fig. 10, p. 88).

Convenient reductive cleavage at the ester group resulting in an almost quantitative yield of (6) and the corresponding C_{25}-alcohol (8) was achieved by use of lithium aluminium hydride. The two products, which represent all of the 45 carbon atoms of flexirubin dimethyl ether, were easily separable on silica gel.

The substitution in benzene ring A (the ring included in the octaenoic acid portion of flexirubin) was determined by isolation of 4-methoxy-3-methylbenzoic acid (9) from the products of ozone degradation of flexirubin dimethyl ether.

Structure elucidation of the C_{20}-phenol proved more difficult, although the nature of the substituents became clear rapidly. In addition, shifts on acetylation of the free -OH group of $\Delta\delta + 0.25$ ppm for one of the aromatic protons, which in (6) (in $CDCl_3$) appeared equivalent, and $\Delta\delta - 0.11$ ppm for the α-methylene group of the dodecyl substituent provided relatively early confirmation for locating the C_{12}-alkyl chain and one proton *ortho* to the hydroxyl group as shown in Fig. 3.

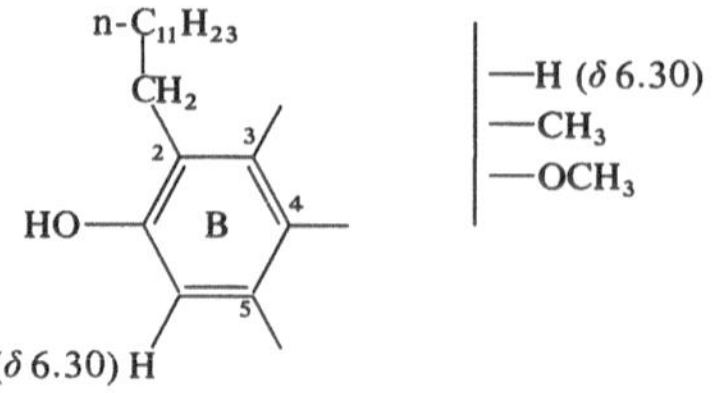

$$n\text{-}C_{11}H_{23}$$
$$\underset{}{CH_2}$$

—H (δ 6.30)
—CH_3
—OCH_3

HO——B——

(δ 6.30) H

Fig. 3. Partial structure of C_{20}-phenol (6) (δ-values in $CDCl_3$)

However, as regards the location of the methoxyl and the methyl group we at first were led astray by the ^{1}H-nmr spectra (90 MHz). The C_{20}-phenol and its derivatives show the signals of two aromatic protons as only weakly broadened singlets although these protons, as transpired later, are *meta* to each other. To provide definite evidence the positional isomers listed in Table 1 were synthetized for comparison of their ^{1}H-nmr spectra. This work later stood us in good stead in the total synthesis of flexirubin dimethyl ether.

References, pp. 109–111

Table 1. *Proton Magnetic Resonance Data (90 MHz) of the Acetyl Esters of the C_{20}-phenol from Flexirubin Dimethyl Ether and of Some Synthetized Isomeric Methoxy- and Methylsubstituted 2-dodecyl Phenols*

Acetyl ester of substituted 2-dodecylphenol:	Resonance signals [ppm]			
	aromatic H	Ar-OCH$_3$	Ar-CH$_2$-	Ar-CH$_3$
3-methoxy-5-methyl-	6.52 (s), 6.45 (s)	3.76	2.46 (t)	2.25
4-methoxy-5-methyl-	6.78 (s), 6.65 (s)	3.80	2.45 (t)	2.17
5-methoxy-3-methyl-	6.63 (s), 6.43 (s)	3.75	2.43 (t)	2.31
5-methoxy-4-methyl-	6.98 (s), 6.50 (s)	3.79	2.43 (t)	2.19
C_{20}-phenol from flexirubin	6.55 (s), 6.44 (s)	3.77	2.45 (t)	2.27

As the result of these experiments flexirubin could be assigned structure (**10**). It consists of an ω-phenyl-hexadecaoctaenecarboxylic acid in ester linkage with a resorcinol; the latter carries a n-dodecyl group at position 2 and a methyl group at C-5.

(**10**)

Formula (**10**) raises the question of the configuration of the polyene in flexirubin. The electronic spectra of *cis*-polyenes show *"cis"*-peaks, which result from steric hindrance in the planar arrangement of the molecule (*20*). In addition, introducing *cis* double bonds causes a decrease in the extinction together with disappearance of the fine structure of the long wavelength maximum. Such effects depend on the position of the *cis* double bond and are especially pronounced if the *cis* double bond is situated in the middle of a long polyene chain. A good example of this effect is seen in the spectrum of 6,8-di-*cis*-flexirubin dimethyl ether (**11**) (Fig. 4, line a), which was obtained in the course of synthetic studies (*21*).

(**11**)

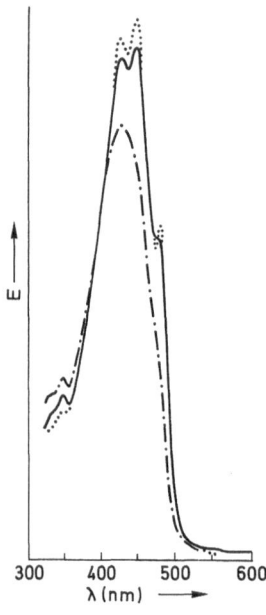

Fig. 4. Electronic spectra of: a) 6,8-di-cis-flexirubin dimethyl ether (**11**) –·–·–; b) flexirubin dimethyl ether (**4**) ······; c) (**4**) after iodine treatment —— (identical with (**11**) after iodine treatment)

Figure 4 also presents the spectra of flexirubin dimethyl ether before and after iodine isomerization: after iodine treatment the extinction value of flexirubin ether decreases because of some *trans-cis* isomerization and a weak *cis* peak appears. This experiment provides no evidence for the presence of any cis double bonds in genuine flexirubin and makes most likely that genuine flexirubin is all-*trans* configurated.

The structure elucidation of the other more than 30 different flexirubin-type pigments which are listed in Table 2 is based exclusively on spectroscopic investigations of a) the original pigments, b) their methylation derivatives, c) their hydrogenated methylation products and d) the alcoholic and phenolic components obtained by reductive cleavage of the hydrogenated permethyl ethers at the ester linkage. These degradations can be carried out on a milligram scale: an example is the structure elucidation of (**12**), a minor pigment from *Fx. elegans*, which was performed with no more than 2 mg (*22*).

$$H_3C \qquad\qquad\qquad n\text{-}H_{25}C_{12} \quad OH$$

$$HO\text{—}\;\text{—[CH}\!=\!\text{CH]}_8\text{—CO}_2\text{—}$$

$$Cl \qquad\qquad\qquad\qquad CH_3$$

(**12**)

References, pp. 109–111

III. Isolation and Separation of Flexirubin-Type Pigments

Fermentation of the microorganisms can be done in small shaking cultures as well as in small or large-size fermentors (12). The medium usually used is Fx A 2.m. (12), which consists of peptone, yeast extract and magnesium sulfate. Typical fermentation times vary from 8 to 20 hours.

Since the pigments are produced and deposited within the cells, the cell material is harvested by centrifugation and – without prior drying – repeatedly extracted with acetone. From the residue of the acetone extracts the raw pigments are separated by extraction with toluene. Yields are rather low and are, for total pigments, in the range of 0.1% of wet cells only.

In contrast to the situation prevailing in *Flexibacter elegans*, where more than 90% of the raw pigment consist of flexirubin (**10**) which can be purified easily by a single thin layer chromatography (TLC), flexirubin-type pigments are usually produced as complicated mixtures of various individual components. An extreme example is represented by the raw pigment from *Cytophaga johnsonae* which consists of some carotenoids (mainly zeaxanthin) and about 25 individual flexirubin-type pigments. Since these pigments do not differ in functional groups and to a major extent are structural homologues, their polarity is quite similar and therefore their separation and isolation can constitute a difficult task.

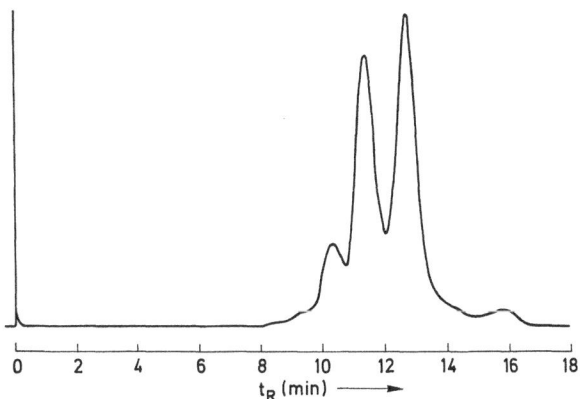

Fig. 5. HPLC-separation of methylated fraction P10-B from the pigment mixture produced by *Cytophaga johnsonae*. (Column: nucleosil 10-C_{18}; eluent: acetone/tetrahydrofuran/water = 74/9/17; detection: UV/VIS at 436 nm)

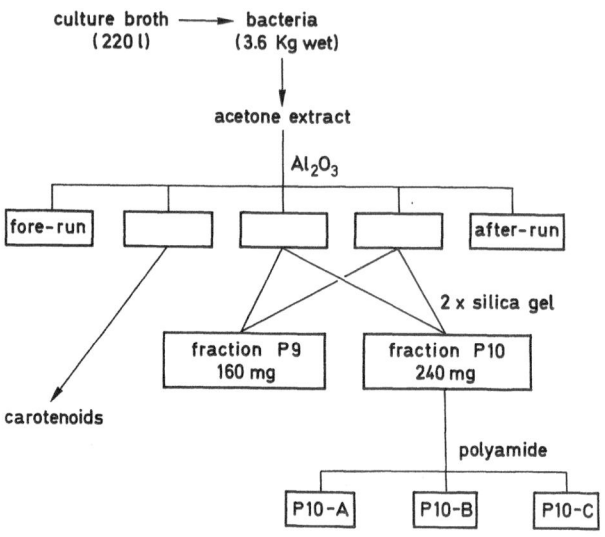

Fig. 6. Chromatographic separation of the raw pigments from *C. johnsonae*

Chromatography on aluminium oxide is useful for separating flexi-rubin-type pigments from carotenoids. Repeated chromatography over silica gel permits separation of chlorinated flexirubin-type pigments from non-chlorinated ones, whereas polyamide is suitable for fractionation of the pigments according to the length of their polyene system. Structural differences depending on only the kind of alkyl substituent at C-2' and/or C-5' need high performance liquid chromatography (HPLC). Good separation results are achieved after methylation of the pigment mixture using reversed phase silica gel C_{18} (*13, 14*). An example is given in Fig. 5.

Suitable combinations of the chromatographic methods mentioned above permitted isolation of the more than 25 individual pigments from *C. johnsonae* (*14*) as shown in Figs. 5 and 6.

IV. General Comments on Structure Elucidation of Flexirubin-Type Pigments

In addition to the electronic spectra from which the polyene length can be deduced, proton magnetic resonance and in particular mass spectrometry contributed importantly to the structure determinations

R^1 OH

HO—⟨ A ⟩—[CH=CH]$_8$—CO$_2$—⟨ B ⟩

R^2

(13) $R^1 = C_{11}H_{23}$ $R^2 = C_5H_{11}$
(14) $R^1 = C_{10}H_{21}$ $R^2 = C_6H_{13}$

(23). ^{13}C-nmr spectrometry was of minor importance mainly because of the small amounts of substances available.

Whereas proton magnetic resonance gives information on the substitution pattern of the aromatic rings and allows facile detection of branching in the alkyl chains, mass spectrometry provides quite a number of structural data (see below) and is of special value if benzene ring B of a flexirubin-type pigment carries two different alkyl substituents (larger than methyl) in the 2'- and 5'-position. Characteristic fragments allow to deduce the exact size of each alkyl substituent and, in addition, it is possible to distinguish the 2'-substituent from the 5'-substituent unambiguously.

Mass spectrometry also served as the major tool for recognition of the presence of mixtures in cases when separation could not be achieved. Thus even by HPLC it was not possible to separate the mixture of isomeric pigments (13) and (14) produced among others from Cytophaga spec. strain Samoa and C. johnsonae (13, 14).

However, from the key fragments in the mass spectrum of the hydrogenated and methylated mixture of the inseparable isomers the structures of both components could be deduced (23) except for the location of the branching points in the alkyl chains which was solved by ^{1}H-nmr spectrometry of the mixture. The structures deduced in this way were corroborated by gc/ms studies of the phenolic fraction obtained by cleavage with lithium aluminium hydride. Using a capillary column the two isomeric phenols containing ring B and its substituents could be separated (13, 23, 24) (Fig. 7). Comparison with the corresponding synthetic dialkylated resorcinol monomethyl ethers proved the deduced structures (24, 25).

It seems worthwhile to mention that lower homologues in the 2,5-dialkyl resorcinol series exhibit antibiotic activity in the plate diffusion test (25).

Table 2 lists all individual flexirubin-type pigments, whose structures have been established up to now and the organisms which produced them.

From Table 2 the general structural features of flexirubin-type pigments can be summarized as follows: the skeleton consists of an ester

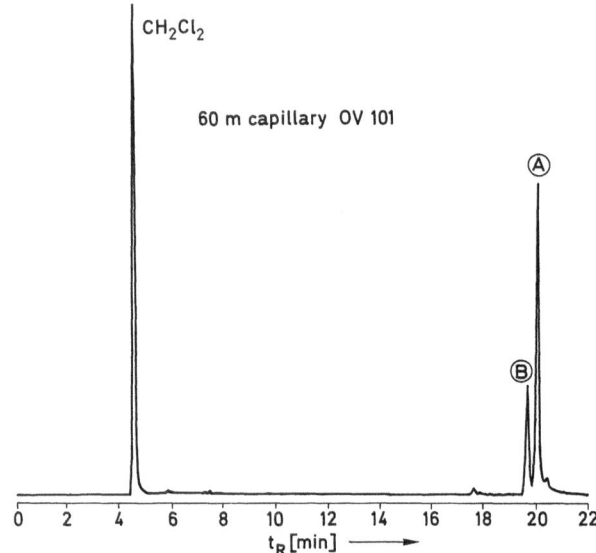

Fig. 7. GC-separation of the phenolic fraction from ester cleavage of the mixture of
(13) and (14)
(Ⓐ: 3-methoxy-2-(9-methyldecyl)-5-pentylphenol;
Ⓑ: 2-decyl-3-methoxy-5-(4-methylpentyl)phenol)

formed from an ω-hydroxyphenyl-substituted polyenoic acid and a 2,5-
dialkylated resorcinol. The polyene system is regularly conjugated and
does not carry any further substituents, in particular no methyl groups.
Structural variability is observed with regard to

a) the length of the polyene, which – according to our present knowledge – ranges from hexaenes to octaenes (n: 6→8).

b) the size of the alkyl substituents in ring B which range for R^1 (at
C-2′) from decyl to tridecyl (R^1: $C_{10}H_{21} \rightarrow C_{13}H_{27}$), and for R^2
(at C-5′) from CH_3, and propyl to hexyl (R^2: CH_3, $C_3H_7 \rightarrow C_6H_{13}$).

c) the substitution of ring A in positions 3 and 5; R^3 can be H, CH_3
or Cl, and R^4 is H or Cl.

The terminal branching frequently observed in R^1 and R^2 is obviously related whether the number of carbon atoms in the alkyl substituent is even or odd. In the case of R^1, an even number of carbon
atoms is always correlated with an unbranched alkyl group whereas
an odd number – except for (15) – indicates terminal branching. For
R^2 the situation is reversed: in that position an even number of carbon
atoms is associated with terminal branching. This empirical rule later

References, pp. 109–111

Table 2. *Structures of Flexirubin-type Pigments Produced by Various Bacteria of the Cytophaga/Flexibacter Group* [a) Fx. elegans, b) Flavobacterium spec. strain C 1/2, c) Cytophaga spec. strain Samoa, d) Cytophaga johnsonae Cy jl]

Structure	Name/ code	n	R^1	R^2	R^3	R^4	Organism	Ref.
(10)	flexirubin	8	$n\text{-}C_{12}H_{25}$	CH_3	CH_3	H	a)	(18)
(12)	chloro-flexirubin	8	$n\text{-}C_{12}H_{25}$	CH_3	CH_3	Cl	a)	(22)
(15)	homo-flexirubin	8	$n\text{-}C_{13}H_{27}$	CH_3	CH_3	H	a)	(34)
(16)	Fla-P5	8	$n\text{-}C_{12}H_{25}$	CH_3	H	H	b)	(17)
(17)	Fla-P4	8	$n\text{-}C_{12}H_{25}$	C_3H_7	H	H	b)	(17)
(18)	Fla-P3	8	$n\text{-}C_{12}H_{25}$	$CH_2CH(CH_3)_2$	H	H	b)	(17)
(19)	P16-1	8	$C_{10}H_{21}*$	$C_5H_{11}*$	H	H	c), d)	(13, 14)
(13)	P16-2-I	8	$[CH_2]_8CH(CH_3)_2$	$n\text{-}C_5H_{11}$	H	H	c), d)	(13, 14)
(14)	P16-2-II	8	$n\text{-}C_{10}H_{21}$	$[CH_2]_3CH(CH_3)_2$	H	H	c), d)	(13, 14)
(20)	P16-3	8	$[CH_2]_8CH(CH_3)_2$	$[CH_2]_3CH(CH_3)_2$	H	H	c), d)	(13, 14)
(21)	P15-1	8	$C_{10}H_{21}*$	$C_5H_{11}*$	Cl	H	c), d)	(13, 14)
(22)	P15-2-I	8	$[CH_2]_8CH(CH_3)_2$	$n\text{-}C_5H_{11}$	Cl	H	c), d)	(13, 14)
(23)	P15-2-II	8	$n\text{-}C_{10}H_{21}$	$[CH_2]_3CH(CH_3)_2$	Cl	H	c), d)	(13, 14)
(24)	P15-3	8	$[CH_2]_8CH(CH_3)_2$	$[CH_2]_3CH(CH_3)_2$	Cl	H	c), d)	(13, 14)
(25)	Fla-P2	8	$n\text{-}C_{12}H_{25}$	$n\text{-}C_3H_7$	Cl	H	b)	(17)
(26)	Fla-P1	8	$n\text{-}C_{12}H_{25}$	$CH_2CH(CH_3)_2$	Cl	H	b)	(17)
(27)	P10-B-1	7	$C_{10}H_{21}*$	$C_5H_{11}*$	H	H	d)	(14)
(28)	P10-A-1	6						(14)
(29)	P10-B-2-I	7	$[CH_2]_8CH(CH_3)_2$	$n\text{-}C_5H_{11}$	H	H	d)	(14)
(30)	P10-A-2-I	6						(14)
(31)	P10-B-2-II	7	$n\text{-}C_{10}H_{21}$	$[CH_2]_3CH(CH_3)_2$	H	H	d)	(14)
(32)	P10-A-2-II	6						(14)
(33)	P10-B-3	7	$[CH_2]_8CH(CH_3)_2$	$[CH_2]_3CH(CH_3)_2$	H	H	d)	(14)
(34)	P10-A-3	6						(14)
(35)	P9-B-1	7	$C_{10}H_{21}*$	$C_5H_{11}*$	Cl	H	d)	(14)
(36)	P9-A-1	6						(14)
(37)	P9-B-2-I	7	$[CH_2]_8CH(CH_3)_2$	$n\text{-}C_5H_{11}$	Cl	H	d)	(14)
(38)	P9-A-2-I	6						(14)
(39)	P9-B-2-II	7	$n\text{-}C_{10}H_{21}$	$[CH_2]_3CH(CH_3)_2$	Cl	H	d)	(14)
(40)	P9-A-2-II	6						(14)
(41)	P9-B-3	7	$[CH_2]_8CH(CH_3)_2$	$[CH_2]_3CH(CH_3)_2$	Cl	H	d)	(14)
(42)	P9-A-3	6						(14)
(43)	P8	7	$[CH_2]_8CH(CH_3)_2$	$[CH_2]_3CH(CH_3)_2$	Cl	Cl	d)	(14)

* Branching not proved.

was explained and corroborated by results from biosynthetic studies
(see p. 107).

V. Spectroscopic Properties of Flexirubin-Type Pigments

1. Mass Spectra

Mass spectrometric investigations were very helpful in work on
the structure of flexirubin-type pigments. In spite of the fact that the
underivatized pigments are suitable for the usual electron-impact mass
spectrometric studies it is more advantageous to use their methylation
products since the lower polarity and higher thermostability of these
derivatives guarantee better reproducibility of the spectra.

Flexirubin-type pigments exhibit a characteristic mass spectrometric
fragmentation pattern from which it is not only possible to recognize
a member of this class of compounds easily, but also to obtain substan-
tial structure information. It is even possible to deduce structural details, which can hardly be determined by any other spectroscopic method. Figure 8 shows the ms of flexirubin (10), its dimethyl ether (4)
and its hexadecahydrodimethyl ether (5).

Characteristic for the basic polyene structure of the pigments is
the loss of 78 mu which is caused by consecutive intramolecular electrocyclic reactions within the polyene part of the molecule and expulsion
of benzene. The mechanism parallels the known mass spectrometric
loss of toluene and xylene from carotenoids (Fig. 9) (11). The exclusive
loss of benzene from the molecular ions of flexirubin-type pigments
demonstrates the non-isoprenoid character of their polyene systems.

The intensity of the M-78-peak is not influenced significantly by
the length of the polyene ($n=6\rightarrow8$), but increases when the sample
is exposed to higher temperatures inside the ion source of the mass
spectrometer over a longer time. A triple bond within the conjugated
system (e.g. (44) (21)) can inhibit the typical ms-induced expulsion
of parts of the unsaturated system.

(44)

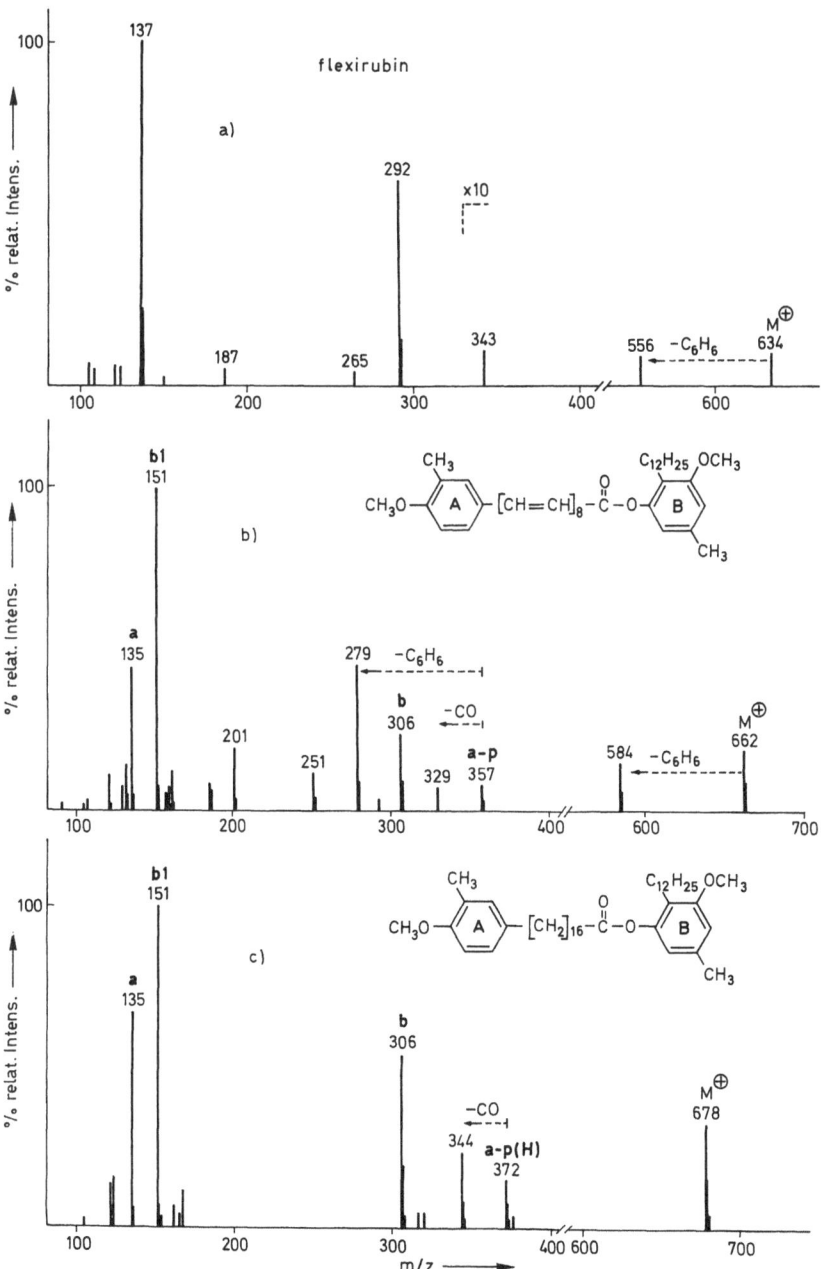

Fig. 8. Mass spectra of a) flexirubin (10), b) its dimethyl ether (4) and c) its hexadecahydro-dimethyl ether (5) (electron impact; 70 eV)

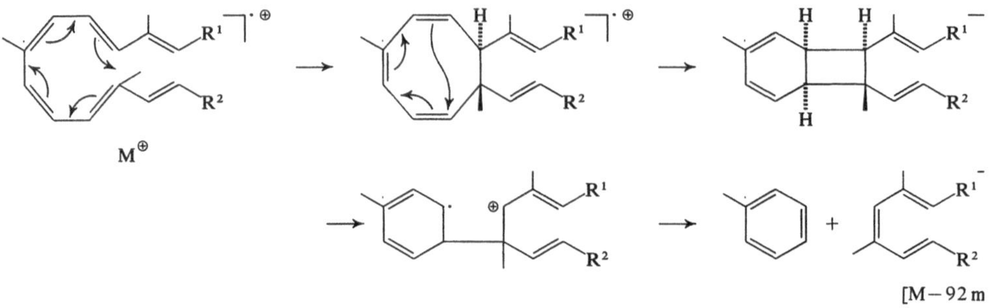

H. Achenbach:

[M−92 m

Fig. 9. Mass spectrometric loss of toluene (and correspondingly xylene) from carotenoids according to (*11*)

The bonds at which the further mass spectrometric breakdown of flexirubin-type pigments occurs preferentially are indicated in Fig. 10.

Primary key fragments occur from breakage at the ester bond. The charge can be located on either fission product and this causes ions **a-p** (ring A + polyene) and **b** (ring B). This process is even more pronounced in the ms of the hydrogenated derivatives. Another presumably secondary fragmentation process produces fragment **a**, which con-

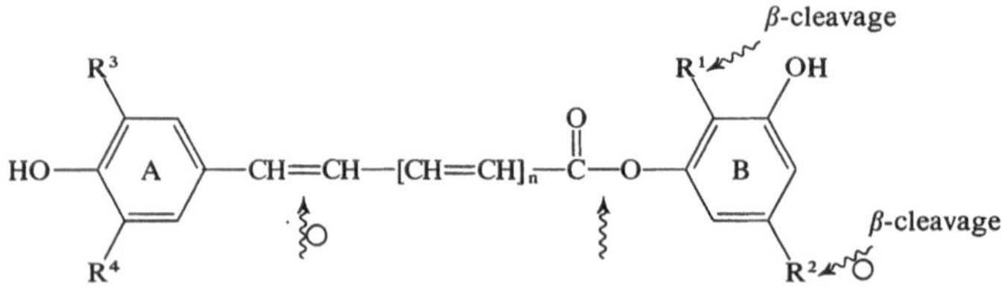

a) Bonds, which are preferably broken in the ms of flexirubin-type pigments

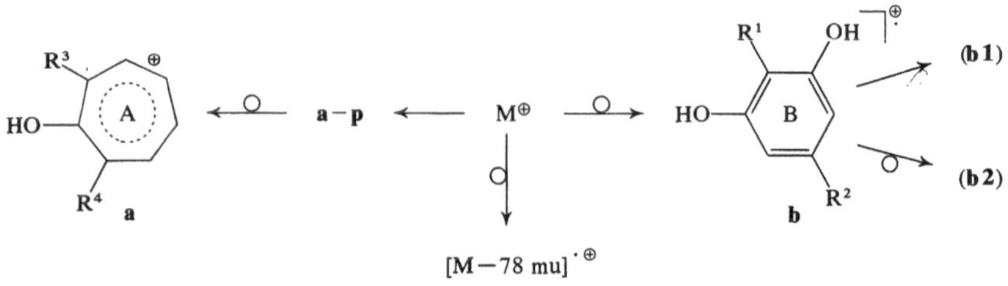

[M−78 mu]⁺

b) Structures of key fragments

Fig. 10. Mass spectrometric breakdown of flexirubin-type pigments: a) bonds, which are preferably broken; b) structures of key fragments

References, pp. 109–111

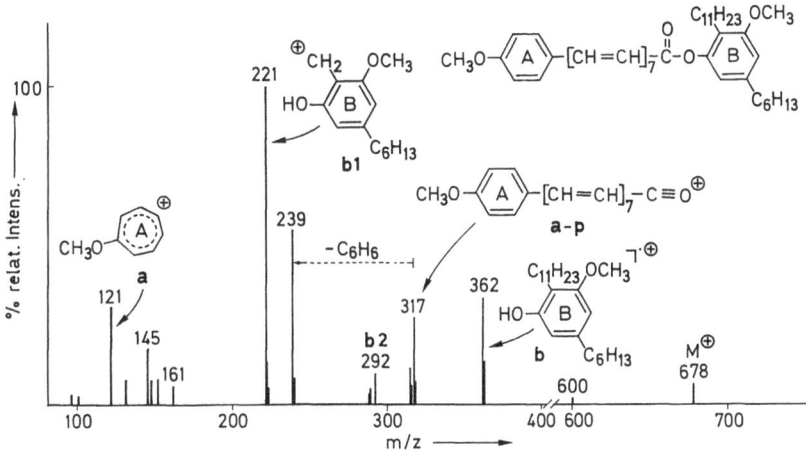

Fig. 11. Mass spectrum of the dimethyl ether of pigment (33) (=P10-B-3) from
C. johnsonae

tains only benzene ring A (together with its substituents) and one CH_2-group which originates from the ω-carbon atom of the polyene chain. Therefore **a** is indicative of the masses of the substituents at ring A and – when it is subtracted from key fragment **a-p** – of the length of the polyene.

Secondary fragmentation at the alkyl substituent R^1 of fragment **b** causes formation of ion **b1**, from which the size of this alkyl substituent can be deduced.

In most flexirubin-type pigments, R^2 is larger than CH_3. In that case R^2 can also split off, giving rise to an ion **b2** in addition to ion **b1** (Fig. 11).

It is noteworthy and analytically useful that the secondary fragmentation processes, which involve the alkyl groups of fragment **b**, are different for R^1 and for R^2. R^1, which is next to two *ortho* substituents always is split by a simple β-cleavage and this results in an uneven mass number for fragment **b1**. In contrast, a McLafferty-rearrangement takes place at R^2, thus leading to an even-numbered fragment **b2** (Fig. 12).

These observations are in agreement with the results of ms-studies of 5-alkylated (26) and 2,5-dialkylated resorcinols and their mono-methyl ethers (23, 25). Therefore, the key fragments **b1** and **b2** permit one not only to deduce the sizes of R^1 and R^2, but also to establish their positions unambiguously.

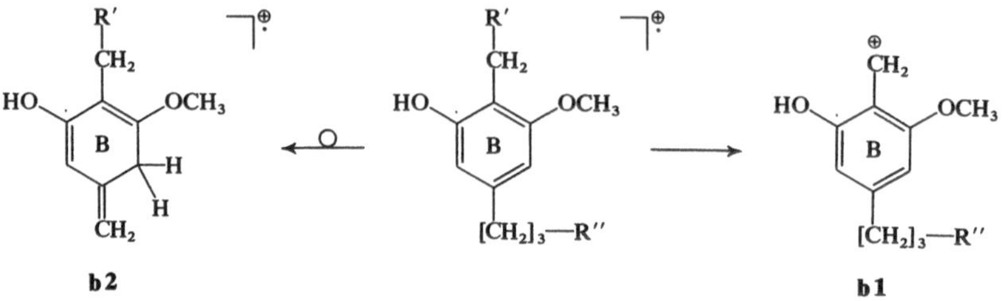

H. Achenbach:

Fig. 12. Formation of key fragments **b1** and **b2**

The following table summarizes the structural information on flexi-rubin-type pigments available from the mass spectra (Table 3):

Table 3. *Structural Information from Key Fragments in the MS of Flexirubin-Type Pigments*

Key fragment	Structural information
M – 78 mu	non-isoprenoid polyene
a-p [or **a-p(H)**] and **a**	length of polyene
a	substitution at ring A
b	substitution at ring B
b1	size of R^1
b2	size of R^2

Of course, chlorine-content is easily recognized from the character-istic pattern of isotope peaks in the molecular ion as well as in the fragment ions.

Mass spectral investigations are also useful for gaining structure information on chromatographic fractions containing mixtures of flexi-rubin-type pigments and to optimize their separation. The mass spec-trum shown in Fig. 13 was taken from an apparent pure column chro-matographic fraction after methylation. It shows that the fraction con-sists of a mixture of three homologous components of the flexirubin-type all possessing the same length of the polyene ($n = 7$) and the same substitution in ring A but differing in the size of the alkyl groups in ring B. From the ratio of intensities of the molecular ions the quanti-tative composition was determined to be approximately 10:4:3. Later this mixture could be separated by HPLC (Fig. 5, p. 81).

By mass spectrometry it was also possible to answer structural ques-tions for mixtures of isomers which resisted all efforts of separation. Fig. 14 presents the mass spectrum of an apparent pure fraction from HPLC, which for better thermostability and fewer mass spectral side fragmentations had been methylated and hydrogenated prior to mass

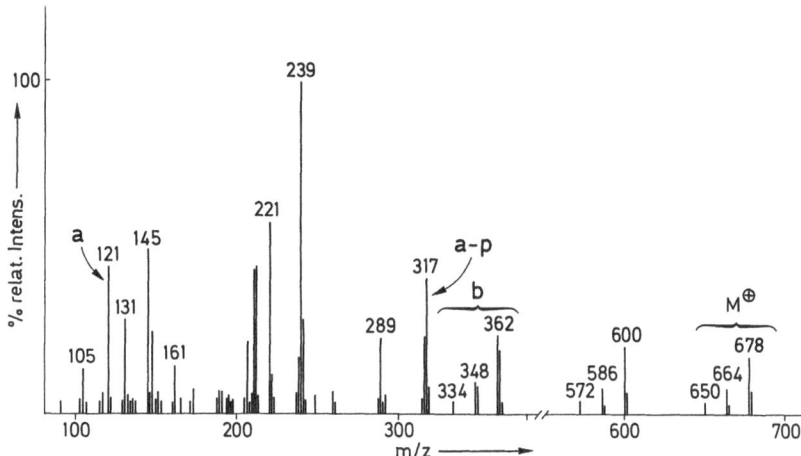

Fig. 13. Mass spectrum of a methylated apparently pure pigment fraction from
C. johnsonae (fraction P10-B)

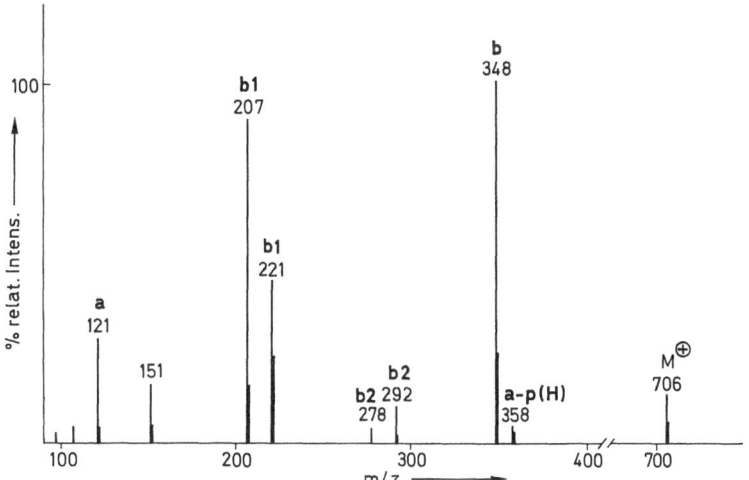

Fig. 14. Mass spectrum of a HPLC-pure fraction from *Cytophaga spec.* strain *Samoa*
(methylated and hydrogenated)

spectral analysis. The spectrum exhibits four secondary fragments of type **b**, two of the **b1**-type and two of the **b2**-type; all other characteristic ions appear only once. The secondary fragments of **b** can be arranged into two sets, which fit structures **(13)** and **(14)**. Therefore, the seemingly pure fraction is an approximately 2:1 mixture of inseparable isomers which are equal in respect to the sum R^1 plus R^2.

2. Proton-NMR-Spectra

The proton resonance spectra of flexirubin-type pigments are essentially trivial. In work on the structures ^{1}H-nmr spectrometry served mainly to establish the substitution pattern of the aromatic rings and to recognize branching in the alkyl groups.

When measured at 90 MHz the *meta*-standing protons at C-4' and C-6' (of ring B) appear as two slightly broadened individual singlets in (**4**) and the hydrogenation product (**5**), but they collapse to one signal in the C_{20}-phenol (**6**). At 180 MHz, in some flexirubin-type pigments an AB-system becomes observable for these protons with $J \sim$ 1 Hz. Their surprisingly small coupling constant might be due to the two oxygen substituents in ring B (*27*).

At 180 MHz the olefinic protons situated at the first and last double bond can be recognized. They exhibit coupling constants in the 12 to 15 Hz range which demonstrate their *trans*-relationship.

The ^{1}H-signals of the ω-positioned isopropyl (or methyl) groups of the alkyl substituents R^1 and R^2 in ring B occur at slightly different fields ($\Delta\delta \sim 0.02$ ppm). This effect is obviously due to the differing influence of the aromatic system on these protons, since they become equivalent when deuterobenzene is used as the solvent for nmr measurements.

3. ^{13}C-NMR-Spectra

The ^{13}C-nmr data of flexirubin dimethyl ether are presented in Fig. 15. The values for its cleavage products (**6**) and (**8**) are compiled in Tables 8 and 9 (page 102, 103).

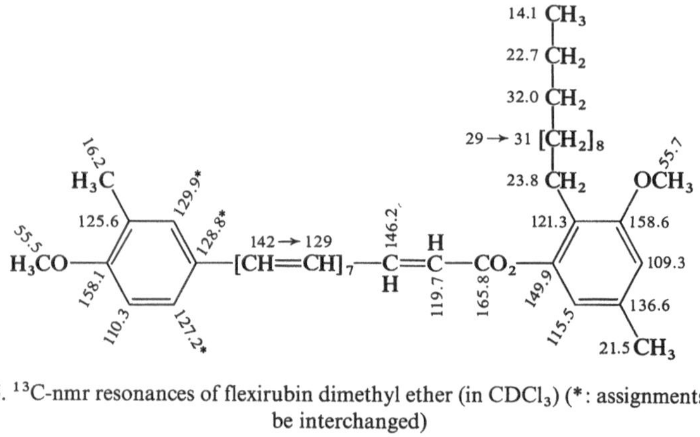

Fig. 15. ^{13}C-nmr resonances of flexirubin dimethyl ether (in CDCl$_3$) (*: assignments might be interchanged)

The assignments of the individual signals are based on increment calculations, on single-frequency off-resonance ^{1}H-decoupling experiments and on ^{13}C-measurements of suitable model compounds (28).

4. Electronic Spectra

From the position of the long wavelength maximum the number of double bonds present in the polyene system can be deduced. A bathochromic shift of about 25 nm occurs immediately on addition of NaOH, because of the free phenolic OH-groups.

Presence of chlorine in ring A decreases the maxima slightly (2 nm) (Table 4). Methylation does not change the spectrum in MeOH significantly, but permits measurement in hexane. In this solvent a fine structure is observed as is also typical for carotenoids and other polyenes (29, 30).

Table 4. *Long Wave Length Maxima of Chlorine-free Flexirubin-Type Pigments*

Flexirubin-type pigment length of polyene	λ_{max} [nm]		Dimethyl derivative fine structure at λ_{max} hexane
	MeOH	MeOH + NaOH	
hexane	417	442	396, 415, 437
heptaene	434	458	412, 432, 460
octaene	448	473	421, 448, 475

VI. Synthetic Approaches to Flexirubin-Type Pigments

Pure flexirubin-type pigments are not easily available. They are produced microbiologically in low yields only, and usually occur in complicated and hard-to-separate mixtures. Therefore, their total synthesis is attractive.

The first target was flexirubin dimethyl ether (4), whose synthesis has been achieved by the strategy depicted in Fig. 16.

This synthetic concept outlines a route which can be applied easily to the preparation of other flexirubin-type pigments by changing only (a) single building block(s). Thereby, it allows a) variation of the length of the polyene, b) variation of substitution at ring A, c) variation of

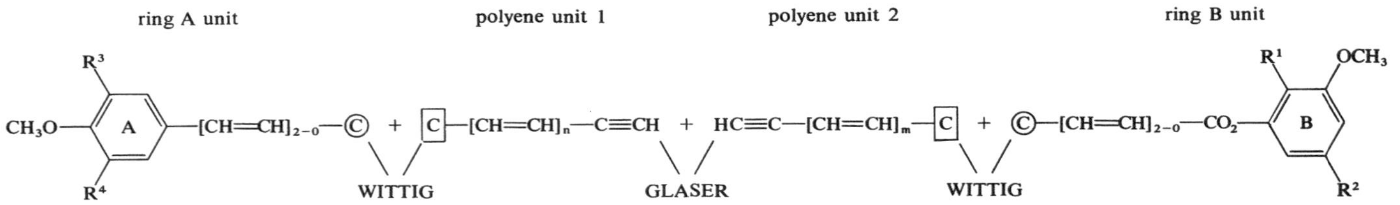

Fig. 16. Concept for the synthesis of flexirubin-type pigments

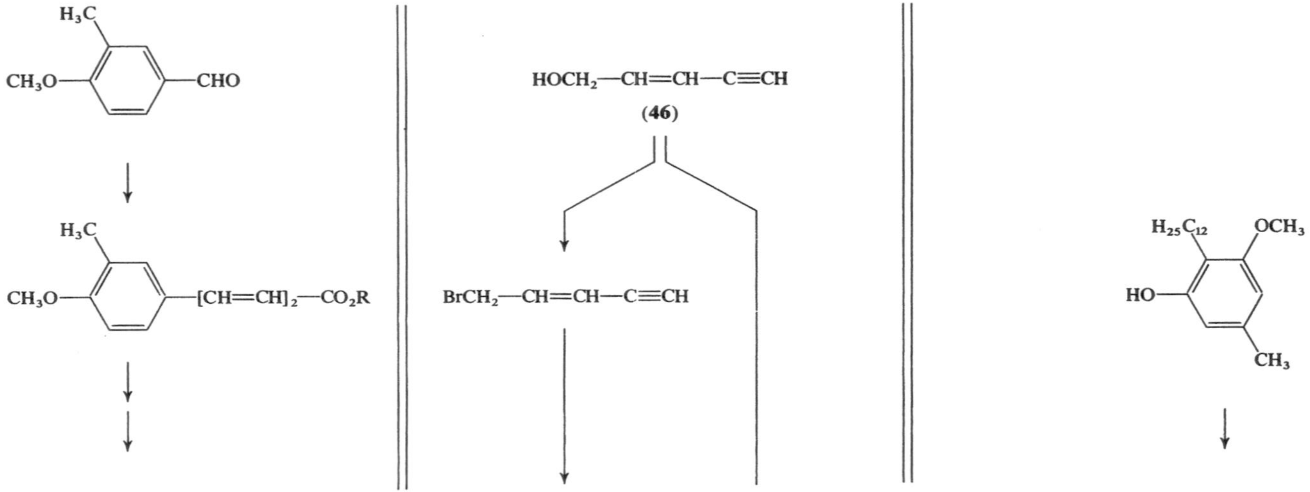

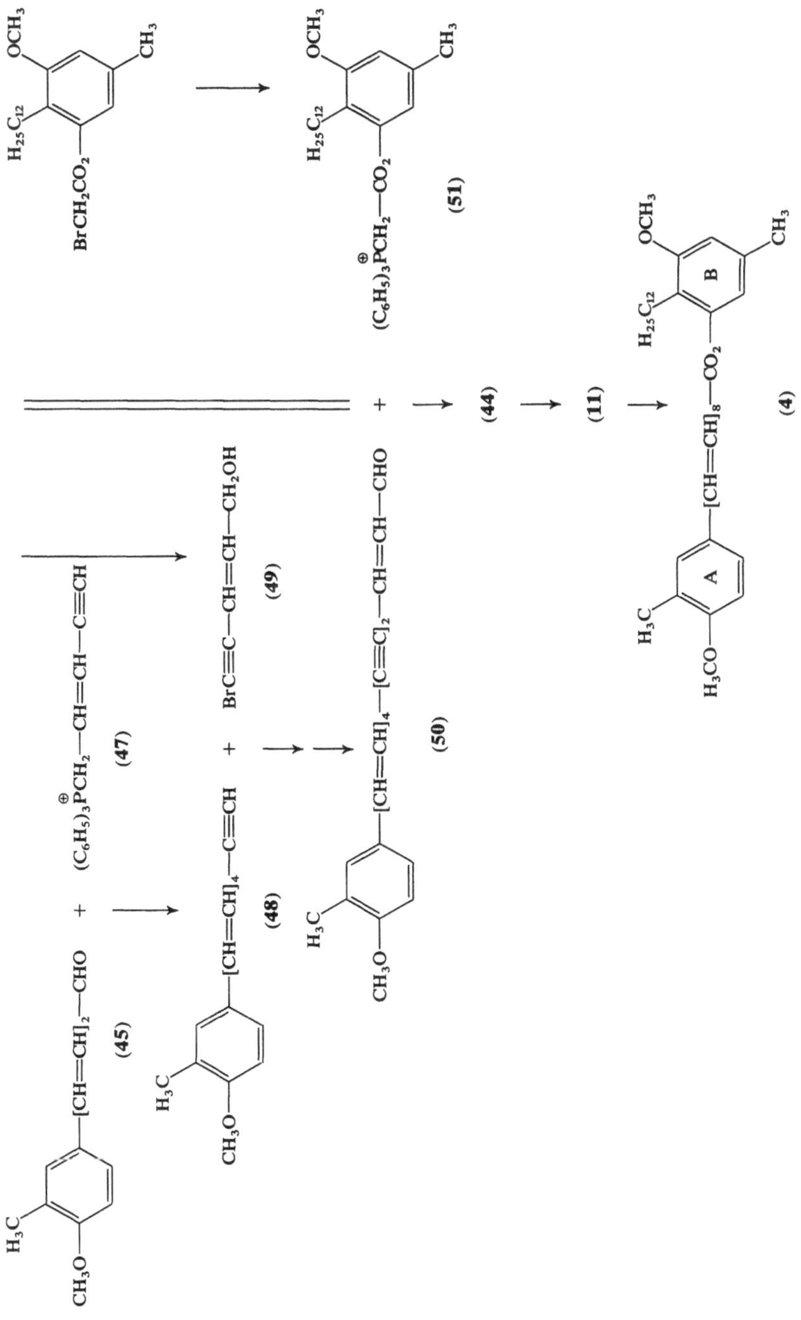

Fig. 17. Synthesis of flexirubin dimethyl ether (4) (21)

substitution at ring B. It also avoids the stage of the free polyenoic acid which is extremely insoluble.

The molecule is put together by combination of two major building blocks containing the aromatic rings A and B, respectively, and of two minor building blocks, from which the central part of the polyene is constructed (polyene unit 1 and polyene unit 2). Because of easier accessibility and better chemical stability the use of two smaller units for the polyene was found to be superior to the use of only one larger unit. Coupling of the building blocks is achieved by two Wittig reactions and one Glaser reaction. The Glaser reaction, which is performed asymmetrically by the modification according to CHODKEWICZ (31) results in a polyene-yne. The triple bonds contribute to the chemical stability of the unsaturated system until they are subjected to partial hydrogenation and subsequent isomerization in the ultimate step of the synthesis (29).

The details of the synthesis of flexirubin dimethyl ether are depicted in Fig. 17.

5-(4-Methoxy-3-methylphenyl)pentadiene-al (45) was used as the building block for the aromatic ring A. As starting material for the preparation of both polyene units, 2-pentene-4-yn-1-ol (46) was used; it can be prepared easily from epichlorohydrin and sodium acetylide according to (32). Reaction with $(C_6H_5)_3PBr_2$ and then $(C_6H_5)_3P$ converts (46) to the phosphonium salt (47) which is coupled with (45) by a Wittig reaction to give the tetraene-yne (48). The second polyene unit (49) is prepared from (46) by treatment with HOBr and added to (48) by a Chodkewicz coupling reaction. Oxidation of the alcoholic group with manganese dioxide produces the pentaene-diyne aldehyde (50) which is needed for the final Wittig reaction with the ring B component (51).

Partial hydrogenation using Lindlar catalyst leads to the 6,8-di-*cis*-stereoisomer of flexirubin dimethyl ether (11) which is isomerized to flexirubin dimethyl ether by treatment with iodine. The isomerization can be followed nicely by electronic spectroscopy (Fig. 4).

In spite of the fact that the various chemical steps are relatively routine, the following observations are pertinent to the preparation of the building block containing ring B and the ester bond: starting with orcinol dimethyl ether, the introduction of the C_{12}-side chain is performed regioselectively at C-2 by addition of dodecyl bromide to the lithiated orcinol derivative, but preparation of the monomethyl ether requires cleavage of the methoxyl groups and partial remethylation. This route was found to be superior to acylation of orcinol or orcinol dimethyl ether with dodecanoyl chloride in the presence of $AlCl_3$ and subsequent reduction of the carbonyl group (by $NaBH_4$

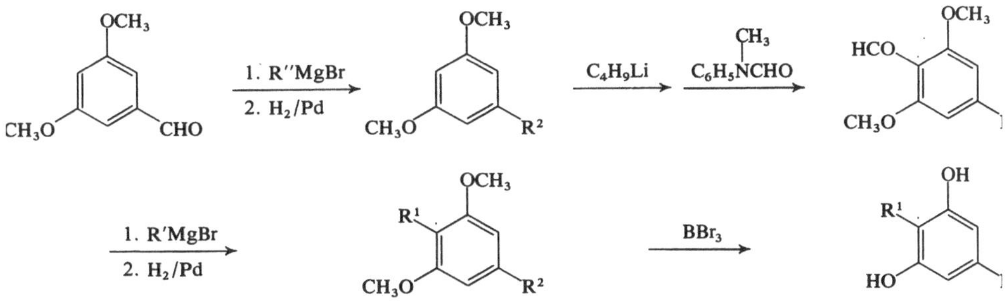

Fig. 18. Synthetic route to 2,5-dialkylated resorcinols

and then H_2/Pd) because the products from the latter sequence were mixtures of 2- and 4-acylated isomers and their separation was a tedious process. This disadvantage was not compensated for by the advantage that the 2-acylated compounds can be converted directly to the mono-ethers either by partial methylation of 2-dodecanoylorcinol or by partial cleavage of 2-dodecanoylorcinol dimethyl ether with BCl_3.

The 2,5-dialkylated resorcinols (52) and (53) ($R^1, R^2 > CH_3$) were prepared from 3,5-dimethoxybenzaldehyde by the reaction sequence outlined in Fig. 18 (25).

VII. Biosynthesis of Flexirubin-Type Pigments

A considerable number of ^{14}C-labelled potential precursors have been added to cultures of *Flexibacter elegans*. The rates of incorporation of radioactivity into flexirubin are summarized in Table 5 (33, 34). These data clearly show that acetic acid acts as the main building block for flexirubin, whereas mevalonic acid does not participate in its biosynthesis (33). Unfortunately, the low rates of incorporation together with the low yields of formation of flexirubin pose great difficulties for biosynthetic ^{13}C-experiments.

More detailed information on the biosynthesis resulted from the chemical degradation of the ^{14}C]flexirubins and location of the incorporated radioactivity. The isolated and purified ^{14}C]flexirubins were first converted into the hexadecahydroflexirubin dimethyl ethers. In order to avoid quenching effects the radioactivity measurements were

Table 5. *Incorporation of Radioactivity into Cell Material and into Flexirubin by Fx. elegans from Various Precursors After 6 to 8 hrs Fermentation Time (33, 28)*

Precursor	Rate of incorporation into:		Precursor	Rate of incorporation into:	
	cells [‰]	flexirubin [‰]		cells [‰]	flexirubin [‰]
[1-^{14}C]acetate*	150	0.7	L-[^{14}CH$_3$]methionine	50	0.35
[2-^{14}C]acetate*,a	110	0.3	L-[U-^{14}C]phenylalanine	44	0.01
[1-^{14}C]butyrate*,a	160	0.2	L-[U-^{14}C]tyrosine	73	0.4
[3-^{14}C]cinnamic acid*,b	16	0.03	DL-[2-^{14}C]tyrosine	56	0.45
D-[U-^{14}C]glucose	290	0.9	L-[1-^{14}C]tyrosine	120	0.86
[2-^{14}C]malonate*	20	0.15			

* As sodium salts; a fermentation time 14 to 16 hrs; b fermentation time 3 hrs.

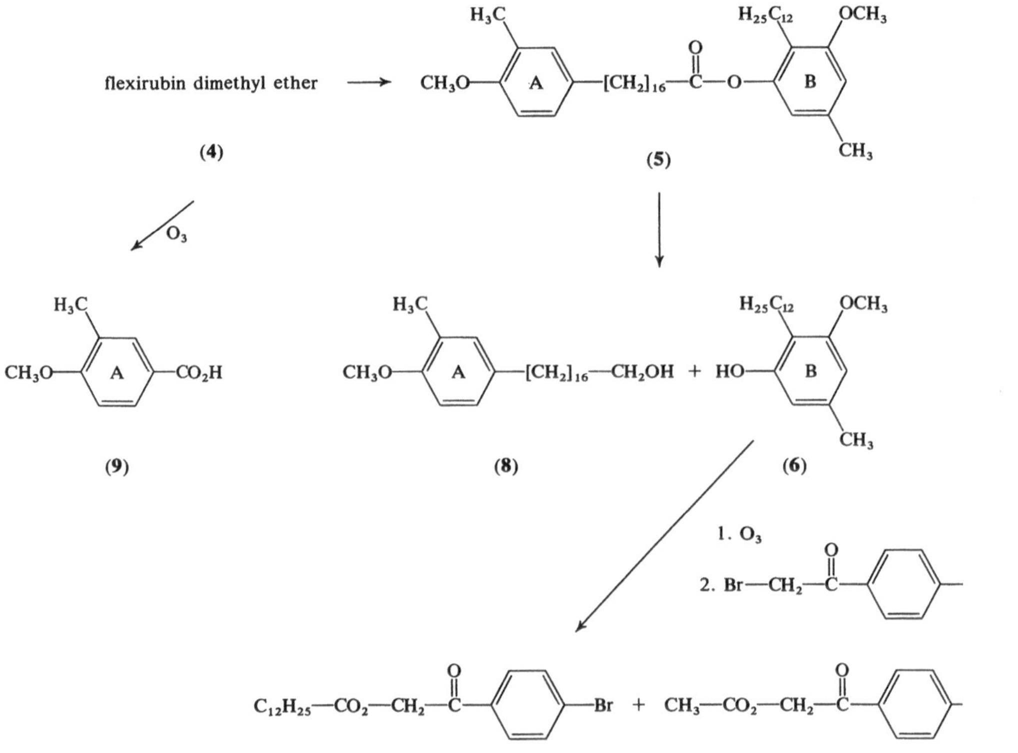

Fig. 19. Chemical degradation of labelled flexirubins

Table 6. *Distribution of Radioactivity in Degradation Products After Administration of Various Precursors*

Precursor	Distribution of radioactivity					
	hexadecahydro-dimethyl ether (5)	C_{25}-alcohol (8)	C_{20}-phenol (6)	(54)	(55)	(9)
[1-^{14}C]acetate	100%	41%	63%	39%	8%	
[2-^{14}C]acetate	100%	40%	62%	38%	8%	
[1-^{14}C]butyrate	100%	32%	67%	43%	7%	
L-[^{14}CH$_3$]methionine	100%	97%	0%			
L-[U-^{14}C]tyrosine	100%	90%	5%			76%
DL-[2-^{14}C]tyrosine	100%	93%	8%			0%

performed on the colourless hydrogenated compounds, which then were cleaved at the ester linkage to the C_{25}-alcohols and the C_{20}-phenols (= resorcinol monomethyl ethers) using lithium aluminium hydride. In addition, the flexirubin dimethyl ethers from the labelled tyrosines were degraded by ozonolysis to the 4-methoxy-3-methyl-benzoic acids (9) (Fig. 19) (*35*).

The results of the radioactivity measurements in the flexirubins and their degradation products are compiled in Table 6 (*34, 35*).

From these results the following conclusions can be drawn:
1) Only the methyl group at ring A derives from methionine.
2) Tyrosine is a precursor of benzene ring A and the last carbon atoms of the polyene chain.
3) The major part of the polyene chain originates from acetate.
4) Among all tested precursors only butyrate and acetate are incorporated into benzene ring B and its substituents at a considerable extent.

The observed ratio of radioactivity between the C_{25}-alcohol and the C_{20}-phenol in the [^{14}C]acetate experiment is in agreement with equal incorporation of acetate into C-1 to C-14 of the polyene chain (= 7 acetate units) and into all carbon atoms of benzene ring B and its substituents (9 to 10 acetate units). Degradation of the C_{20}-phenol from the [^{14}C]acetate experiments by ozone (Fig. 19) and radioactivity measurements on the resulting *p*-bromophenacyl esters of tridecanoic acid (54, from the dodecyl substituent) and of acetic acid (55, from the methyl substituent) corroborate this interpretation (Table 7) (*34*).

These results suggest a polyketide pathway for the octaenoic acid part of flexirubin – with *p*-hydroxycinnamic acid (from tyrosine) as the starter unit – and for ring B and its alkyl substituents as well.

Table 7. *Distribution of Radioactivity from [^{14}C]acetates and [^{14}C]butyrate Within Ring B and Its Substituents*

Precursor	Distribution of radioactivity		
	C$_{20}$-phenol (6)	(54)	(55)
[1-^{14}C]acetate	100%	62%	13%
[2-^{14}C]acetate	100%	61%	13%
[1-^{14}C]butyrate	100%	64%	10%

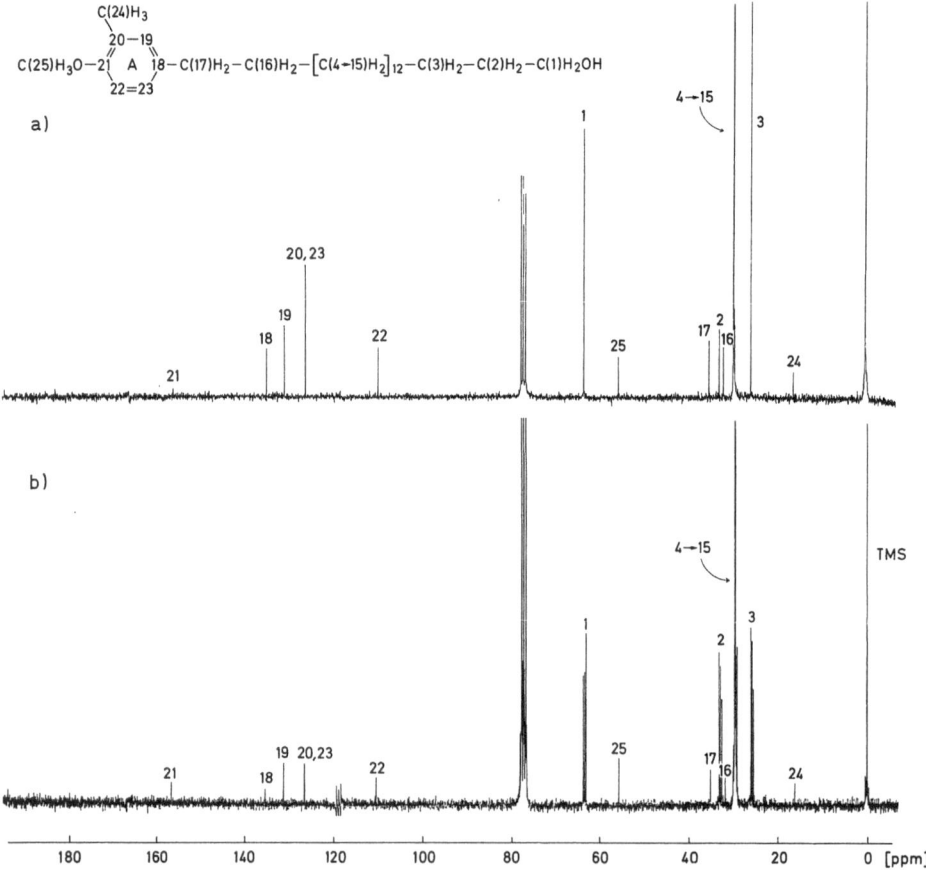

Fig. 20. ^{13}C-nmr spectrum of C$_{25}$-alcohol from [^{13}C]flexirubin produced in the presence of a) [1-^{13}C]acetate and b) [1,2-^{13}C$_2$]acetate

Full proof for a biosynthetic pathway of ring B *via* an orsellinic acid intermediate and some additional information were given by feeding experiments with $[1,2-^{13}C_2]$acetate and $[1-^{13}C]$acetate (*34*). To achieve this, numerous preliminary cultivation and incorporation experiments with $[^{14}C]$acetate had to be carried out to optimize isotope incorporation and flexirubin production. To simplify the interpretation, the ^{13}C-labelled flexirubins were methylated, hydrogenated and cleaved at the ester bond by LiAlH$_4$. The ^{13}C-nmr spectra of the degradation products are shown in Figs. 20 and 21.

Assignments of the individual resonances and their alterations in the labelled substances are compiled in Tables 8 and 9.

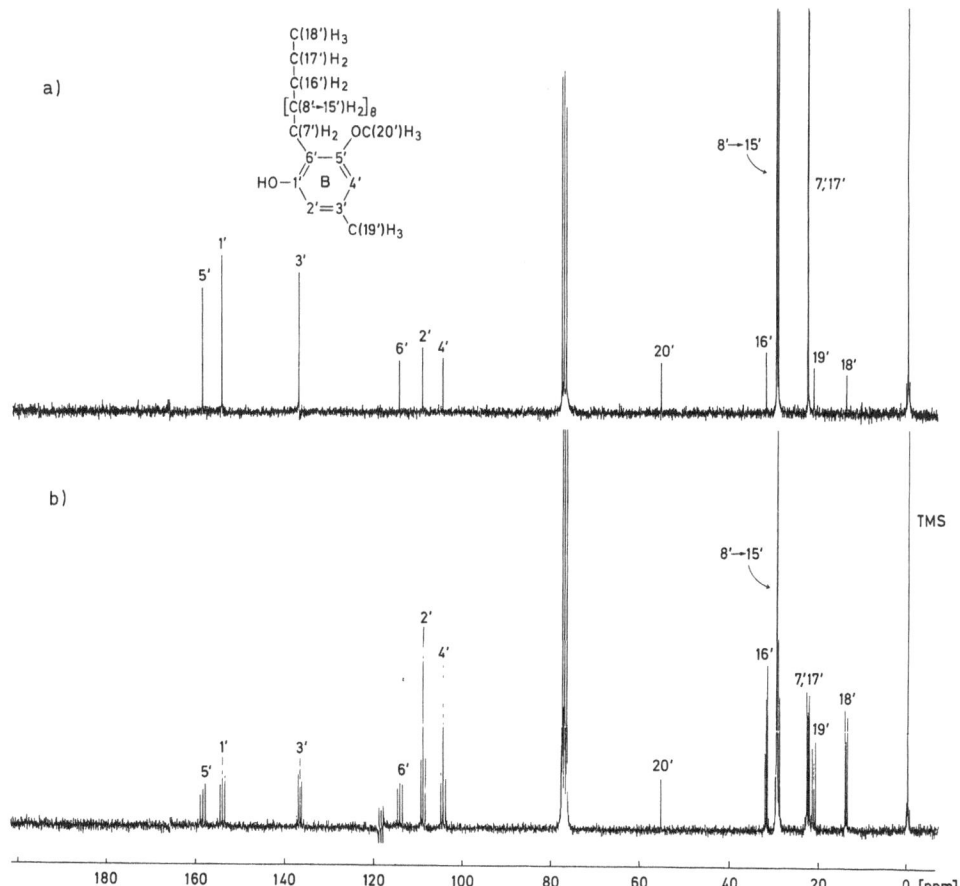

Fig. 21. ^{13}C-nmr spectrum of C$_{20}$-phenol from $[^{13}C]$flexirubin produced in the presence of a) $[1-^{13}C]$acetate and b) $[1,2-^{13}C_2]$acetate

Table 8. *Alteration of ^{13}C signals of C_{25}-alcohol* (**8**) *After Administration of ^{13}C-labelled Acetates*

C-Atom*	^{13}C-Resonance	Observed alteration of signal after feeding with:	
	[δ]	[1-^{13}C]acetate	[1,2-^{13}C$_2$]acetate
1	63.2	enhancement	triplet
2	33.0	none	triplet
3	25.9	enhancement	triplet
4–15	29.1–30.1	partial enhancement	partial triplets
16	31.9	none	none
17	35.2	none	none
18	134.9	none	none
19	131.0	none	none
20	126.5	none	none
21	156.1	none	none
22	110.1	none	none
23	126.5	none	none
24	16.2	none	none
25	55.5	none	none

* Special numbering according to Fig. 20.

With respect to ring A, the results confirm that acetate does not serve as a precursor for ring A and the last carbon atoms of the polyene-oic acid.

With respect to ring B, it is of particular interest to consider the incorporation of ^{13}C from [1,2-^{13}C$_2$]acetate into carbons 2′ and 4′*. The signals of both of these atoms show the same pattern, clearly different from all other signals in the spectrum as each is composed of a doublet *and* an enhanced singlet. Whereas the doublet originates from incorporation of a whole C_2-unit from acetate, the enhancement of the singlet must be caused by incorporation of only one carbon atom from acetate with loss of the adjacent carbon atom. As the comparison with the corresponding signals in the spectrum of the C_{20}-phenol from the [1-^{13}C]acetate experiment demonstrates, incorporation of C-2 from acetate and loss of the carboxyl group must be responsible for the enhanced singlets.

This observation and the other data from Table 9 are in accordance with biosynthetic formation of benzene ring B via orsellinic and 3-dode-cylorsellinic acid according to Fig. 22 with loss of the carboxyl group at either stage of the biosynthesis.

Whereas carbon atoms 1 and 5 in orsellinic or 3-substituted orsel-linic acids are non-equivalent they become magnetically equivalent car-

* Special numbering according to Fig. 21.

Table 9. *Alterations of ^{13}C signals of C_{20}-phenol (6) After Administration of ^{13}C-labelled Acetates*

C-Atom*	^{13}C-Resonance	Observed alteration of signal after feeding with:	
	[δ]	[1-^{13}C]acetate	[1,2-^{13}C$_2$]acetate
1'	154.3	enhancement	triplet
2'	109.1	none	triplet **
3'	136.9	enhancement	triplet
4'	104.6	none	triplet **
5'	158.7	enhancement	triplet
6'	114.4	none	triplet
7'	23.0	enhancement	triplet
8'–15'	29.1–30.0	partial enhancement	partial triplets
16'	32.0	none	triplet
17'	22.7	enhancement	triplet
18'	14.1	none	triplet
19'	21.5	none	triplet
20'	55.7	none	none

* Special numbering according to Fig. 21.
** In addition the central signal is significantly enhanced.

bon atoms on decarboxylation to the corresponding orcinols as long as the orcinol is not converted to an unsymmetrical derivative. In case of a [1,2-^{13}C$_2$]acetate experiment this means that the ^{13}C signal of C-2'/C-4'* in an – symmetrical – orcinol, which biosynthetically originates *via* orsellinic acid, consists partly of an enhanced singlet (since the signal of C-(1) in the orsellinic acid precursor collapses from a doublet to a singlet on loss of the adjacent carboxyl group) and partly of a doublet (from C-(5) in the orsellinic acid precursor, which indeed originates from incorporation of the CH$_3$-group of an intact acetate unit into the aromatic ring).

From the equal ^{13}C patterns of carbon atoms 2'* and 4'* in flexirubin and the C$_{20}$-phenol it can be concluded that biosynthetic decarboxylation of orsellinic acid (or 3-dodecylorsellinic acid) must result in a symmetrical molecule and therefore occurs *before* the ester bond with the polyene unit or its precursor is formed. Obviously, the two major parts of the flexirubin molecule are formed independently and they are linked to an ester at an ultimate stage of the flexirubin biosynthesis.

The results of competition experiments using [^{14}C]acetate in the presence of a) inactive orsellinic acid (57) and b) inactive 3-dodecylorsellinic acid (58) are summarized in Table 10 (*34*).

* Special numbering according to Fig. 21.

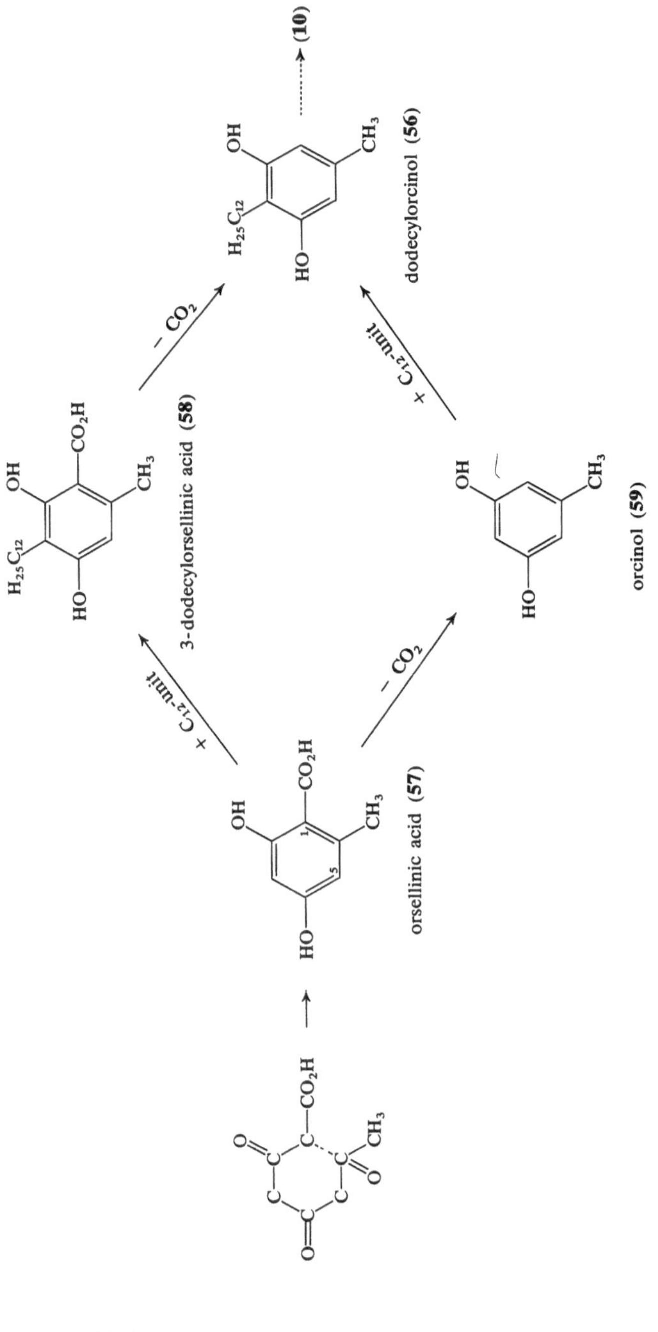

H. Achenbach:

Fig. 22. Possible biosynthetic pathways to dodecylorcinol (**56**) (= ring B and its substituents) *via* orsellinic acid (**57**)

Table 10. *Distribution of Radioactivity in Degradation Products of Flexirubin from Cultures with [1-^{14}C]acetate in the Presence of Orsellinic Acid* (**57**) *and 3-dodecylorsellinic Acid* (**58**)

Precursor	Distribution of radioactivity				
	Hexadecahydro-dimethyl ether (**5**)	C$_{25}$-Alcohol (**8**)	C$_{20}$-Phenol (**6**)	(**54**)	(**55**)
[1-^{14}C]acetate	100%	41%	63% (=100%)	(62%)	(13%)
[1-^{14}C]acetate + (**57**)	100%	48%	49% (=100%)	(86%)	(3%)
[1-^{14}C]acetate + (**58**)	100%	75%	24% (=100%)	(56%)	(11%)

These data show that both inactive compounds characteristically influence the incorporation of acetate into the C$_{20}$-phenol of flexirubin and therefore support the biogenetic scheme outlined in Fig. 22. In the presence of 3-dodecylorsellinic acid (**58**), incorporation of acetate into (**6**) is suppressed and the acetate is found mainly in the C$_{25}$-alcohol (**8**). In agreement with this competing situation is the observation that the distribution of the remaining radioactivity within (**6**) was found to be similar to that observed in the experiment without addition of 3-dodecylorsellinic acid. By contrast, orsellinic acid (**57**) does not influence the ratio of radioactivity between the C$_{25}$-alcohol and the C$_{20}$-phenol so much, but it has a significant effect on the distribution of radioactivity within the C$_{20}$-phenol where under these conditions the radioactivity is enhanced in the C$_{12}$-substituent.

That orcinol (**59**) can be definitely excluded as an intermediate in flexirubin biosynthesis is demonstrated by a study of the metabolites produced by a colourless mutant of *Fx. elegans* in which flexirubin production was blocked. When this mutant was grown in the presence of [2-^{14}C]acetat 3-dodecylorsellinic acid and 2-dodecylorcinol were found to be among the strongly radioactive metabolites and orsellinic acid was also radioactive to a lower extent, whereas orcinol did not contain measurable radioactivity.

Therefore, the pathway of flexirubin biosynthesis can be depicted according to Fig. 23.

No experiments have been performed to decide whether the methyl group at the *m*-position of ring A is introduced from methionine during the biosynthesis of the polyenoic acid or at a very late stage after the ester bond has already been formed.

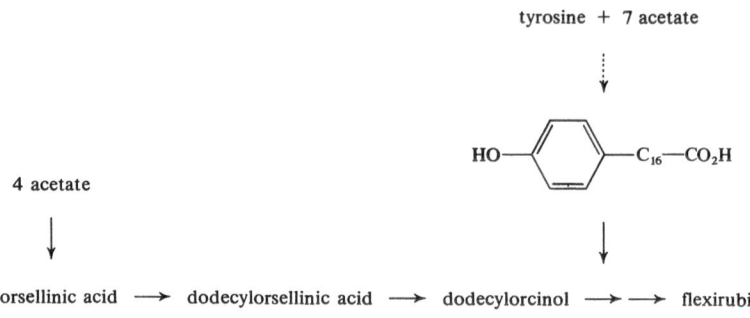

Fig. 23. Biosynthesis of flexirubin

It seems worthwhile to mention briefly the role of propionic acid in flexirubin biosynthesis. Preliminary tests carried out in search for potential biosynthetic precursors of flexirubin revealed a good incorporation of radioactivity from [1-^{14}C]propionate. The results from the experiments with ^{14}C- and ^{13}C-labelled acetate strongly suggested that incorporation of C-1 from propionate into flexirubin might occur by transcarboxylation reactions via acetate. However, this explanation was excluded by degradation experiments which surprisingly demonstrated that more than 90% of the incorporated radioactivity from [1-^{14}C]propionate is located in the tridecanoic acid p-bromophenacyl ester (54) (obtained from the alkyl substituent of ring B according to Fig. 19).

As further investigations proved, (54) contained a very small amount of tetradecanoic acid p-bromophenacyl ester. This higher homologue of (54) is present in a concentration of approximately 0.8% of the C_{13}-acid ester and is therefore not detected in routine degradation experiments. However, purification by HPLC of the p-bromophenacyl ester isolated from ozonolysis revealed that the highly purified C_{13}-acid ester is inactive and the C_{14}-acid derivative carries all the radioactivity from [1-^{14}C]propionate (28, 34).

Thus, propionate acts as the starter molecule for the biosynthesis of a linear C_{13}-side chain in ring B (via 3-tridecylsellinic acid) in the hitherto unknown homoflexirubin (15).

(15) R = n-$C_{13}H_{27}$

Similar concentrations of the C_{14}-acid were also found in the C_{13}-acids obtained by chemical degradation of flexirubin from fermentation experiments without addition of propionate. Therefore, the hitherto unknown pigment (15) which accompanies flexirubin and which is biosynthetized by incorporation of propionate appears to be a normal, but very minor component of the standard pigment-mixture produced by *Fx. elegans.*

The biosynthetic pathway to flexirubin outlined in Fig. 23 can be generalized to all other known structures of flexirubin-type pigments as follows:

a) Ring A and polyenoic acid: Ring A originates from tyrosine which including its three non-cyclic carbon atoms serves as precursor of the starter unit for the condensation of acetate to generate the ω-phenylpolyenoic acid. According to our present knowledge the number of acetate units participating in the biosynthesis of this part of the molecule varies from five to seven. Chlorination can occur at the *meta*-position(s) of ring A and is a late biosynthetic process. Up to now, biosynthetic methylation of ring A is restricted to pigments from *Fx. elegans.* No methylation (or incorporation of isoprene units) has been observed for the polyene system.

b) Ring B and its alkyl substituents: The linear precursor forming ring B is not necessarily built up from four acetate units as is the case for orsellinic acid (57) itself. Longer polyketides also can be precursors for ring B, thus forming higher orsellinic acid homologues (R at $C-6 > CH_3$) as intermediates. The starter unit for the biosynthesis of the linear polyketide precursor of an orsellinic acid homologue is not necessarily acetate (or malonate); isovaleric acid (from leucine) can also serve as the starter and this explains the endbranching of alkyl substituents at C-5′ not uncommonly observed in flexirubin-type pigments.

The same considerations of a fatty acid like biosynthesis with activated acetic or isovaleric acid as the starter can be applied to the linear polyketide, which alkylates (or acylates) the intermediate orsellinic acid (or its higher homologue), thus introducing the alkyl substituent R^1 at C-2′ of flexirubin-type pigments.

It is known, that branched starting units in fatty acid biosynthesis can be generated by degradation of suitable amino acids (36). This was demonstrated for R at C-5′ of flexirubin-type pigments by feeding experiments with [U-^{14}C]leucine and other amino acids to cultures of *Flavobacterium spec.* strain C1/2, which produces a mixture of flexirubin-type pigments with linear and end-branched alkyl substituents at C-5′ (17). Significant enrichment of radioactivity from [U-^{14}C]leucine within this alkyl substituent was accompanied by end branching (37).

The biosynthetic route outlined above well explains the hypothetical rule postulated earlier (*13*). In the structure of flexirubin-type pigments an alkyl substituent at C-5′ ($=R^2$) is always branched, if it contains an even number of carbon atoms, whereas terminal branching of the alkyl at C-2′ ($=R^1$) is associated with an uneven number of carbon atoms (except in homoflexirubin (**15**)).

The biosynthetic situation was used by REICHENBACH *et al.* to suggest an easy test in the search for new flexirubin-type pigments (*38*). The test allows one to distinguish unambiguously flexirubin-type pigments from carotenoids on thin layer chromatographic plates, which is not possible solely on the basis of colour. For this test the organisms are cultivated in the presence of a) radioactive tyrosine and b) radioactive mevalonic acid, and the colouring matters produced are checked by TLC separation and subsequent autoradiography. Incorporation of ^{14}C-tyrosine indicates a flexirubin-type pigment, whereas incorporation of mevalonic acid is characteristic for carotenoids.

VIII. Closing Remarks

According to our present knowledge, the occurrence of flexirubin-type pigments is restricted to bacteria of the *Cytophaga/Flexibacter*-group and among that group, these pigments have been detected preferentially in organisms which were isolated from soil or fresh water. They seem to be absent in most *Cytophaga/Flexibacter*-like bacteria from marine environments (*15*).

The chemotaxonomic utility of flexirubin-type pigments has been demonstrated, in a practical way, during identification of a psychrophilic bacterium isolated from a glacier, whose classification presented difficulties. Since the mass spectrum of the crude pigment mixture from this bacterium exhibited great similarity to that of the pigment complex from *Cytophaga johnsonae*, it was concluded that the microorganism might be closely related. In response to these chemical results, a renewed detailed microbiological study of the organism confirmed the suggestion based on chemotaxonomy (*39*).

The physiological function of flexirubin-type pigments is still obscure. In contrast to carotenoids, these pigments obviously do not serve as photo-protectors. Experiments have proved that a) the production of flexirubin-type pigments is not stimulated by light, and b) that unpigmented mutants grow well under normal laboratory conditions and even in the presence of high light intensities (*15*). The production of flexirubin-type pigments is strictly correlated with the growth of the cells. Resting cells do not produce flexirubin-type pigments nor is there growth of cells without pigment biosynthesis (*1, 15*).

It should also be pointed out that in many bacteria of the *Cytophaga/Flexibacter* group flexirubin-type pigments are produced besides carotenoids. However, it could be demonstrated that pigments of the flexirubin-type are located in the outer membrane of the bacterial cell wall, whereas carotenoids, if additionally present, are found in the cytoplasmatic (or inner) membrane (*40*). This observation might be of major importance with regard to the still unknown physiological function of the novel flexirubin-type pigments.

References

1. REICHENBACH, H., H. KLEINIG, and H. ACHENBACH: The Pigments of *Flexibacter elegans*: Novel and Chemosystematically Useful Compounds. Arch. Microbiol. **101**, 131 (1974).

2. REICHENBACH, H.: Private communication.

3. STACKEBRANDT, E.: Das hierarchische System der Eubakterien: Problem- und Lösungsansätze. Forum Mikrobiol. **9**, 255 (1986).

4. REICHENBACH H., and O.B. WEEKS: The Flavobacterium–Cytophaga Group. Verlag Chemie, Weinheim-Deerfield Beach (Florida)-Basel 1981.

5. MCMEEKIN,T.A., J.T. PATTERSON, and J.G. MURRAY: An Initial Approach to the Taxonomy of Some Gram-Negative Yellow Rods. J. Appl. Bact. **34**, 699 (1971).

6. REICHENBACH, H., and H. KLEINIG: Die Carotinoide der Myxobakterien. Zbl. Bakt., I. Abt. Orig. A **220**, 458 (1972).

7. AASEN, A.J., and S. LIAAEN JENSEN: Carotenoids of Flexibacteria. III. The Structures of Flexixanthin and Deoxyflexixanthin. Acta Chem. Scand. **20**, 1970 (1966).

8. AASEN, A.J., and S. LIAAEN JENSEN: Carotenoids of Flexibacteria. IV. The Carotenoids of two Further Pigment Types. Acta Chem. Scand. **20**, 2322 (1966).

9. LEWIN, R.A., and D.M. LOUNSBERY: Isolation, Cultivation and Characterization of *Flexibacteria*. J. Gen. Microbiol. **58**, 145 (1969).

10. SIMON, G.D., and D. WHITE: Growth and Morphological Characteristics of a Species of *Flexibacter*. Arch. Microbiol. **78**, 1 (1971).

11. VETTER, W., G. ENGLERT, N. RIGASSI, and W. SCHWIETER: Spectroscopic Methods. E. Mass Spectrometry. In: Carotenoids, p. 243 (ISLER, O., ed.), Birkhäuser Verlag, Basel 1981.

12. ACHENBACH, H., W. KOHL, W. WACHTER, and H. REICHENBACH: Investigations of the Pigments from *Cytophaga johnsonae Cy jl* – New Flexirubin-Type Pigments. Arch. Microbiol. **117**, 253 (1978).

13. ACHENBACH, H., W. KOHL, S. ALEXANIAN, and H. REICHENBACH: Stoffwechselprodukte von Mikroorganismen XVII: Neue Pigmente vom Flexirubin-Typ aus *Cytophaga spec.* Stamm *Samoa*. Chem. Ber. **112**, 196 (1979).

14. ACHENBACH, H., W. KOHL, and H. REICHENBACH: Stoffwechselprodukte von Mikroorganismen XX: Die Konstitutionen der Pigmente vom Flexirubin-Typ aus *Cytophaga johnsonae Cy jl*. Chem. Ber. **112**, 1999 (1979).

15. REICHENBACH, H., W. KOHL, and H. ACHENBACH: The Flexirubin-Type Pigments, Chemosystematically Useful Compounds. In: The Flavobacterium-Cytophaga Group, p. 101 (REICHENBACH, H., and O.B. WEEDS, ed.), Verlag Chemie, 1981.

16. REICHENBACH, H., W. KOHL, A. BÖTTGER-VETTER, and H. ACHENBACH: Flexirubin-Type Pigments in a Flavobacterium. Arch. Microbiol. **126**, 291 (1980).

17. ACHENBACH, H., W. KOHL, A. BÖTTGER-VETTER, and H. REICHENBACH: Stoffwechsel-produkte von Mikroorganismen XXII: Untersuchung der Pigmente aus *Flavobacterium spec.* Stamm *C1/2.* Tetrahedron **37**, 559 (1981).

18. ACHENBACH, H., W. KOHL, and H. REICHENBACH: Stoffwechselprodukte von Mikroorganismen XI: Flexirubin, ein neuartiges Pigment aus *Flexibacter elegans.* Chem. Ber. **109**, 2490 (1976).

19. GRIPENBERG, J.: Fungus Pigments II. Cortisalin, a New Polyethenoid Pigment. Acta Chem. Scand. **6**, 580 (1952).

20. WEEDON, B.C.L.: Spectroscopic Methods for Élucidating the Structures of Carotenoids. In: Fortschritte der Chemie organischer Naturstoffe **27**, 81 (ZECHMEISTER, L., ed.), Springer-Verlag, Wien-New York 1969.

21. ACHENBACH, H., and J. WITZKE: Totalsynthese des Flexirubindimethylethers. Angew. Chem. **89**, 198 (1977); Angew. Chem. Int. Ed. Engl. **16**, 191 (1977).

22. ACHENBACH, H., W. KOHL, and H. REICHENBACH: 5-Chlorflexirubin, ein Nebenpigment aus *Flexibacter elegans.* Liebigs Ann. Chem. **1977**, 1.

23. ACHENBACH, H., and W. KOHL: Zur Konstitutionsaufklärung der Pigmente vom Flexirubin-Typ – Massenspektrometrische Untersuchungen. Chem. Ber. **112**, 209 (1979).

24. ACHENBACH, H.: The Flexirubin-Type Pigments – A Novel Class of Natural Pigments from Gliding Bacteria. Rev. Latinoamer. Quim. **9**, 111 (1978).

25. ACHENBACH, H., W. KOHL, and B. KUNZE: Über die Synthese einiger antibiotisch wirksamer 2,5-Dialkylresorcine – Synthese der Antibiotika DB 2073 und Stemphol. Chem. Ber. **112**, 1841 (1979).

26. OCCOLOWITZ, J.L.: Mass Spectrometry of Naturally Occuring Alkenyl Phenols and their Derivatives. Anal. Chem. **36**, 2177 (1964).

27. READ, J.M., JR., R.W. CRECELY, R.S. BUTLER, J.E. LOEMKER, and J.H. GOLDSTEIN: Additive Proton-Proton-Coupling Effects in Disubstituted Benzenes. Tetrahedron Lett. **1968**, 1215.

28. BÖTTGER-VETTER, A.: Untersuchungen zur Biogenese der Pigmente vom Flexirubin-Typ. Dissertation, Freiburg 1981.

29. BOHLMANN, F., and H.-J. MANNHARDT: Konstitution und Lichtabsorption, VIII. Mitteil.: Darstellung und Lichtabsorption von Dimethyl-polyenen. Chem. Ber. **89**, 1307 (1956).

30. STERN, E.S., and C.J. TIMMONS: GILLAM and STERN's Introduction to Electronic Absorption Spectroscopy in Organic Chemistry, 3rd ed., p. 75, 188. Edward Arnold, London 1970.

31. CHODKIEWICZ, W., J.S. ALHUWALIA, P. CADIOT, and A. WILLEMART: Preparation of Aliphatic Bifunctional Compounds. Compt. rend. **245**, 322 (1957).

32. HAYNES, L.J., I.M. HEILBRON, E.R.H. JONES, and F. SONDHEIMER: The Reaction between epichlorohydrin and sodium acetylide. A Novel Route to the Ethynyl-ethylenic Alcohol, 2-Penten-4-yn-1-ol. J. Chem. Soc. **1947**, 1583.

33. FAUTZ, E., and H. REICHENBACH: Biosynthesis of Flexirubin: Incorporation of Precursors by the Bacterium *Flexibacter elegans.* Phytochemistry **18**, 957 (1979).

34. ACHENBACH, H., A. BÖTTGER-VETTER, D. HUNKLER, E. FAUTZ, and H. REICHENBACH: Investigations on the Biosynthesis of Flexirubin – The Origin of Benzene Ring B and its Substituents. Tetrahedron **39**, 175 (1983).

35. ACHENBACH, H., A. BÖTTGER-VETTER, E. FAUTZ, and H. REICHENBACH: Untersuchungen zur Biogenese des Flexirubins – Herkunft des Benzolringes A und der aromatischen C-Methylgruppen. Phytochemistry **18**, 961 (1979).

36. KANEDA, T.: Biosynthesis of Branched Chain Fatty Acids II. Microbial Synthesis of Branched Long Chain Fatty Acids from Certain Short Chain Fatty Acid Substrates. J. Biol. Chem. **238**, 1229 (1963).

37. ACHENBACH, H., A. BÖTTGER-VETTER, E. FAUTZ, and H. REICHENBACH: On the Origin of the Branched Alkyl Substituents on Ring B of Flexirubin-Type Pigments. Arch. Microbiol. **132**, 241 (1982).
38. FAUTZ, E., and H. REICHENBACH: A Simple Test for Flexirubin-Type Pigments. FEMS Microbiology Letters **8**, 87 (1980).
39. ARPIN, N., H. REICHENBACH, and H. ACHENBACH: Unpublished results.
40. IRSCHIK, H., and H. REICHENBACH: Intracellular Location of Flexirubins in *Flexibacter elegans* (Cytophagales). Biochim. Biophys. Acta **510**, 1 (1978).

(Received January 27, 1987)

Structure, Stability and Color Variation of Natural Anthocyanins

T. Goto, Laboratory of Organic Chemistry, Faculty of Agriculture, Nagoya University, Chikusa, Nagoya, Japan

With 36 Figures

Contents

1. Introduction

Anthocyanin is the general name applied to the glycosides of anthocyanidin chromophores which are the origin of the red, violet, and blue colors found throughout the plant kingdom, such as the colors of petals, leaves and fruits. Only a few anthocyanidin nuclei have been found in spite of the great variety of plant colors. The major anthocyanidins found in nature are pelargonidin, cyanidin, peonidin, delphinidin,

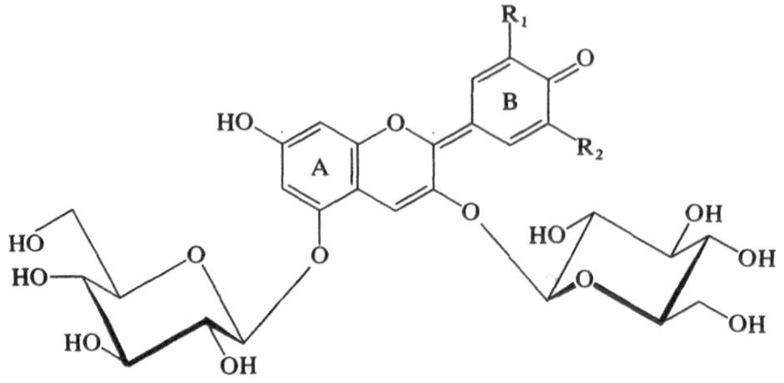

	aglycon
(1) $R_1 = R_2 = H$	Pelargonin (pelargonidin)
(2) $R_1 = OH$ $R_2 = H$	Cyanin (cyanidin)
(3) $R_1 = OCH_3$ $R_2 = H$	Peonin (peonidin)
(4) $R_1 = R_2 = OH$	Delphin (delphinidin)
(5) $R_1 = OCH_3$ $R_2 = OH$	Petunin (petunidin)
(6) $R_1 = R_2 = OCH_3$	Malvin (malvidin)

Fig. 1. 3,5-Diglucosides of common anthocyanins (anhydro base form). The name of the anthocyanidin is in parentheses. [From T. Goto et al.: Ann. New York Acad. Sci. **471**, 155 (1986), with permission]

petunidin and malvidin (Fig. 1). The most common anthocyanins are 3-glucosides and 3,5-diglucosides of the anthocyanidins, but galactose, rhamnose, xylose and arabinose residues are also found. Many anthocyanins containing acylated sugar moieties are also known. The acyl groups are mostly derivatives of cinnamic acid such as p-coumaric, caffeic, and ferulic acid, but include some aliphatic acids such as malonic, succinic, and acetic acid as well.

Anthocyanins change their color with pH; in strongly acidic media they are orange to red, whereas a reddish violet to violet color appears in weakly acidic or neutral solutions (18). A blue color can be produced only in alkaline solution. As the number of hydroxyl group(s) on the B-ring is increased, the visible absorption maximum of the anthocyanidin is shifted to longer wave length; for example, λ max in 0.01% HCl-MeOH solution are: pelargonidin 520 nm, cyanidin 535 nm, peonidin 532 nm, delphinidin 546 nm, petunidin 543 nm and malvidin

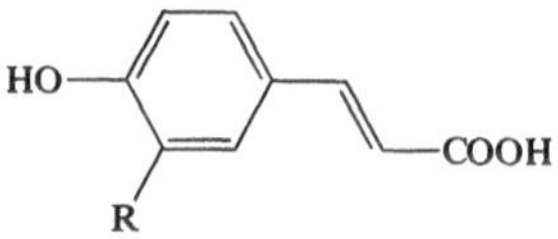

R = H   Coumaric acid
R = OH   Caffeic acid
R = OCH_3   Ferulic acid

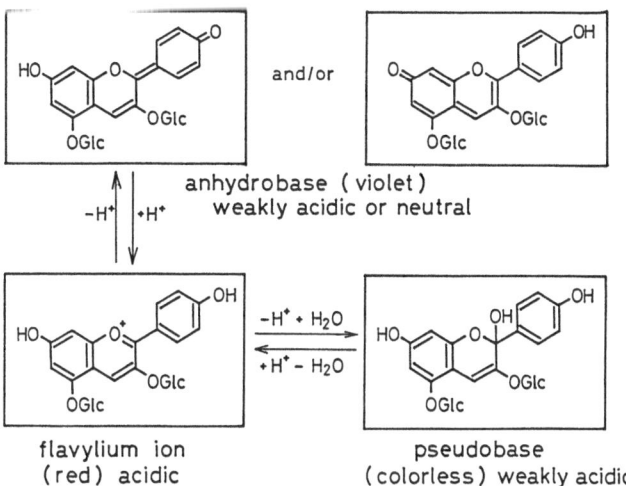

anhydrobase (violet)
weakly acidic or neutral

flavylium ion
(red) acidic

pseudobase
(colorless) weakly acidic

Fig. 2. Structure change of anthocyanin with pH in aqueous solution. [From T. GOTO et al.: Ann. New York Acad. Sci. **471**, 155 (1986), with permission]

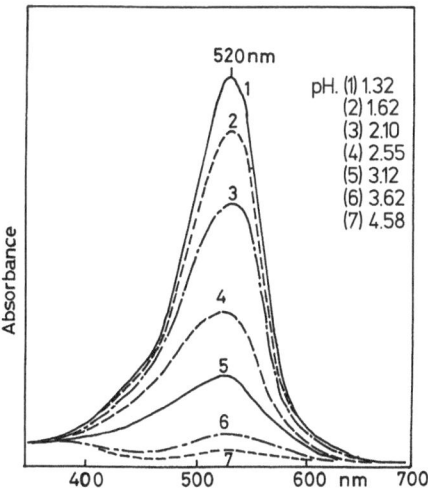

Fig. 3. Visible spectra of awobanin (**12**) in citrate buffer solutions of various pH at 10 minutes after dissolution (10 mm cell)

542 nm (*40*). Anthocyanins are stable in strongly acidic media in the form of a red flavylium ion, but when an acidic solution of an anthocyanin is neutralized to pH 4 to 7, a violet anhydro base is formed first (*112*) but the color disappears rapidly on the result of hydration

with formation of a pseudobase (*18, 20, 21, 22, 45, 64, 65, 75, 109*) (Fig. 2 and 3).

The structural change of anthocyanidins with pH has been extensively studied by BROUILLARD *et al.* (*20, 21, 23*) who concluded that the colorless pseudobase is produced by hydration of a flavylium ion and not from the anhydro base contrary to what had been supposed previously. As flower cell sap is usually weakly acidic, anthocyanins cannot produce stable color, especially blue color, *in vivo* unless there is a mechanism for stabilization and color variation. In this article such mechanisms which have been discovered recently are presented.

2. Structure Determination

Although a huge number of natural anthocyanins have been reported in the literature, their structures are more or less ambiguous with respect to the anomeric configuration and ring size of the sugar moieties, the way the components are connected, particularly the points of attachment of the acyl group(s) in the sugar moieties of acylated anthocyanins, and the geometrical configuration of the double bond when cinnamic acid derivatives are the acyl moieties, since the structures have generally been deduced by acid hydrolysis or oxidative degradation to the components anthocyanidin, sugar(s), and, in the case of acylated anthocyanins, organic acid(s).

This chapter deals with the determination of complete structure and stereochemistry of complex anthocyanins mainly by means of mass and NMR spectrometries. For structures of anthocyanins suggested by conventional methods, earlier reviews should be consulted (*54, 108, 110, 111*).

2.1 Mass Spectra of Anthocyanins

Accurate molecular weight determination by mass spectrometry is very important for structure elucidation of complex anthocyanins, since small components (*16*) that show no NMR signals might be overlooked if the molecular weight is not known. Anthocyanin flavylium salts are, however, highly polar compounds and do not readily show a molecular ion in their EI mass spectra. The recent development of field desorption (FD) ion sources has, however, made it possible to observe the molecular ions of natural anthocyanins; the first example (1982)

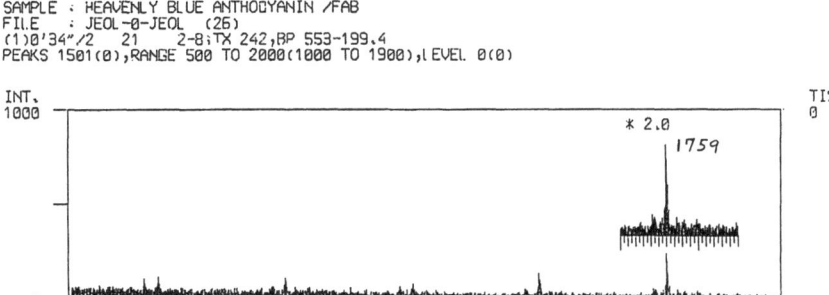

MASS SPECTRUM
SAMPLE : HEAVENLY BLUE ANTHOCYANIN /FAB
FILE : JEOL-0-JEOL (26)
(1)0'34"/2   21   2-8;TX 242,BP 553-199.4
PEAKS 1501(0),RANGE 500 TO 2000(1000 TO 1900),LEVEL 0(0)

Fig. 4. Fast atom bombardment mass spectrum of heavenly blue anthocyanin (**9**)

was that of gentiodelphin (**7**) [m/z 1114 (M+1)] (*33*). FABMS was then found best for obtaining the molecular ion of complex anthocyanins as reported in 1983 by SAITO *et al.* (*89*) (exact mass) and GOTO *et al.* (*34*) in the case of platyconin (**8**) (m/z 1421 M⁺). Peaks corresponding to the flavylium cations of the following acylated anthocyanins have been observed: heavenly blue anthocyanin (**9**) (m/z 1759) (*68*) (Fig. 4), cinerarin (**10**) (m/z 1523) (*36*), malonylawobanin (**11**) (m/z 859) (*35*), and succinylcyanin (**13**) (m/z 711) (*105*). High resolution FABMS (exact mass) (*89*) is a powerful tool for structure analysis of anthocyanins as shown in the case of cyanidin 3-malonylglucoside (**14**) (*16*).

2.2 ¹H-NMR Spectra

Structure analysis of natural anthocyanins by ¹H-NMR spectroscopy should have been very promising, but had not been described in the literature until recently because of the difficulty in obtaining analyzable spectra. Only ¹H-NMR spectra of some synthetic anthocyanidins (aglycons of anthocyanins) had been reported (*78*) until in 1978 GOTO *et al.* (*37*) succeeded in obtaining very fine ¹H-NMR spectra of natural anthocyanins, which made it possible to determine the complete structure and stereochemistry of complex anthocyanins. Thus, the ¹H-NMR spectrum of awobanin (**12**), isolated from commelinin by chromatography on an Avicel column, showed at room temperature quite broad signals that were difficult to analyze, but elevation of the probe temperature resulted in a beautiful spectrum. This disclosed that (1) the *p*-coumaryl group exists completely in the *trans* form, (2) the acyl group is attached to the 6-position of a glucose moiety, and (3) two glucose moieties have a β-pyranoside configuration. Since it had been determined by H_2O_2 oxidation that the acylated glucose is attached at the 3-position, the complete stereostructure of awobanin (**12**) was assigned as 3-0-(6-0-*trans-p*-coumaryl-β-D-glucosyl)-5-0-(β-D-glucosyl)delphinidin (*37, 70*). The Structures of violanin (**15**) (*112*) and shisonin (**16**) were determined similarly (*37, 70*). This was the first instance in which the complete structure and stereochemistry of a naturally-occurring acylated anthocyanin had been determined unambiguously.

That sharp signals in the ^{1}H-NMR spectra of natural anthocyanins have been difficult to obtain might be due to several factors. The pigments are very polar and difficult to dissolve in aprotic solvents, and since hydrochloric acid and cellulose powder chromatography are used for extraction and purification, paramagnetic ions such as ferric ions, which cause broadening of the signals, are liable to contaminate the sample. Careful avoidance of contamination by paramagnetic substances has now made it possible to obtain analyzable ^{1}H-NMR spectra of natural anthocyanins even below room temperature (see Chap. 2.3). ^{13}C-NMR spectra have also been reported (*16, 32*). As an example, the structure determination of one of the most complex natural anthocyanins, heavenly blue anthocyanin (**9**), is outlined below.

2.3 Structure Determination of Heavenly Blue Anthocyanin
(*30, 31, 32, 68*)

Heavenly blue anthocyanin (HBA) is the anthocyanin isolated from the blue petals of a morning glory, *Ipomoea tricolor* (*60*). ISHIKURA and SHIMIZU (*60, 61*) reported that HBA was peonidin 3-diglucoside-5-glucoside acylated with two molecules of caffeic acid, which was further modified by ASEN *et al.* (*8*, see also *58, 82*) to peonidin 3-(dicaffeylisophoroside)-5-glucoside. However, analysis of the NMR and mass spectra, and the hydrolysis products to be discussed later caused GOTO *et al.* (*32*) to revise the composition of HBA to a molecule of peonidin, six molecules of glucose and three molecules of caffeic acid; the molecular weight being 1759 (flavylium cation) as determined by FABMS (*68*) (Fig. 4).

Alkaline hydrolysis of HBA gave tris-deacyl HBA (**18**) (mol. wt. 787, determined by FD-MS) which consists of peonidin and three molecules of glucose. Its structure was suggested by ASEN *et al.* (*8*) to be peonidin 3-sophoroside-5-glucoside on the basis of paper chromatography and UV analyses of the acid hydrolysis products; the anomeric

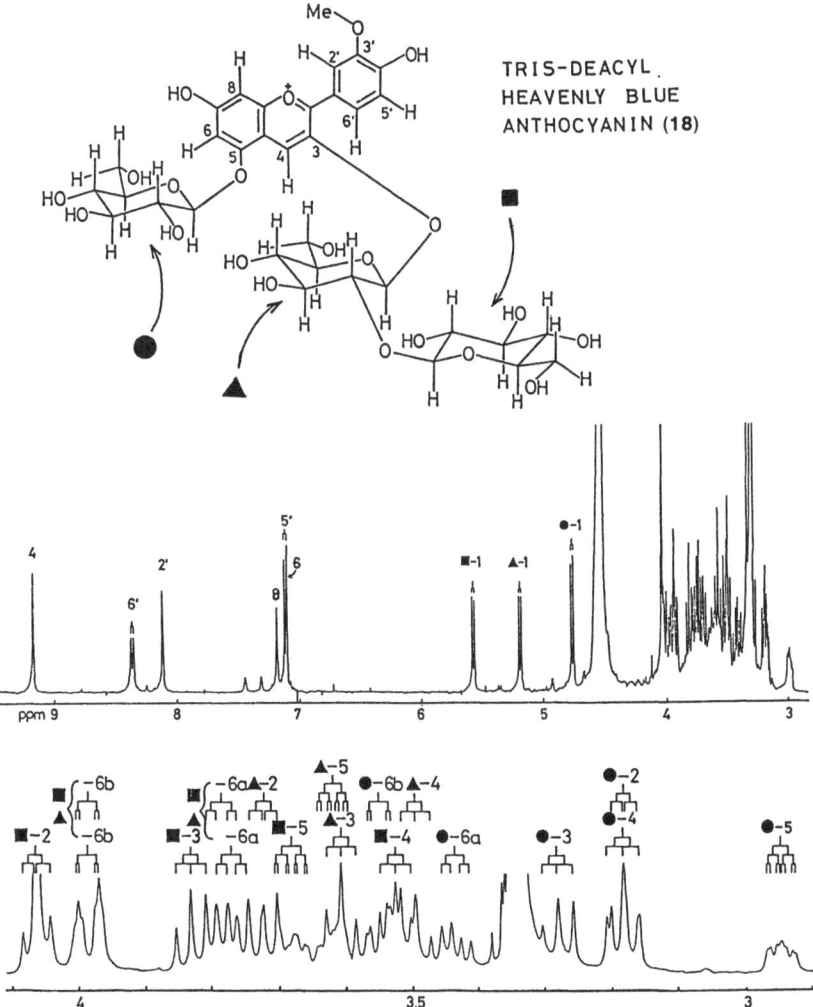

Fig. 5. ^{1}H-NMR spectrum of tris-deacyl HBA (18) at 400 MHz in CD_3OD containing 0.1% DCl at room temperature. [From T. GOTO et al.: Chemistry Lett. 883 (1981), with permission]

configurations have remained undetermined. The complete structure was deduced by analysis of its ^{1}H-NMR spectrum (Fig. 5) (32).

Signals of the three anomeric protons of (18) were assigned by NOE measurements. Signals of three CH_2OH groups were differentiated from other signals by the partially relaxed FT (PRFT) NMR

method as shown in Fig. 6. Other signals of the sugar part were correlated to each other by proton decoupling experiments. Since the $J_{2,3}$'s,
$J_{3,4}$'s and $J_{4,5}$'s of the three glucose moieties are 9 Hz and all of
the $J_{1,2}$'s are 8 Hz, the three sugar moieties must be β-glucosides in
the chair conformation.

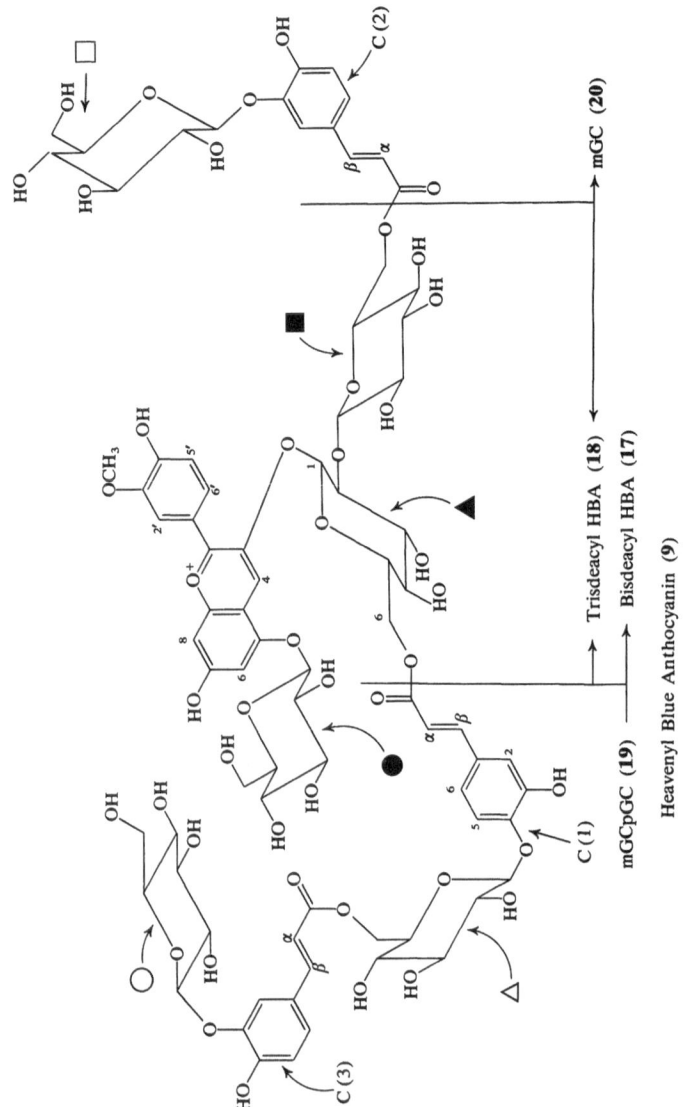

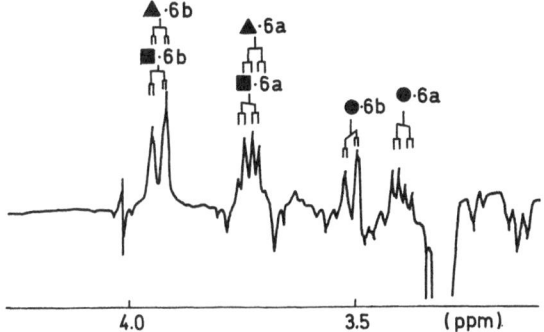

Fig. 6. Partially relaxed FT ¹H-NMR spectrum of tris-deacyl HBA (**18**). Three methylene groups of the glucose units are differenciated from other signals (conditions are same to those in Fig. 5). [From T. Goto *et al.*: Chemistry Lett. 883 (1981), with permission]

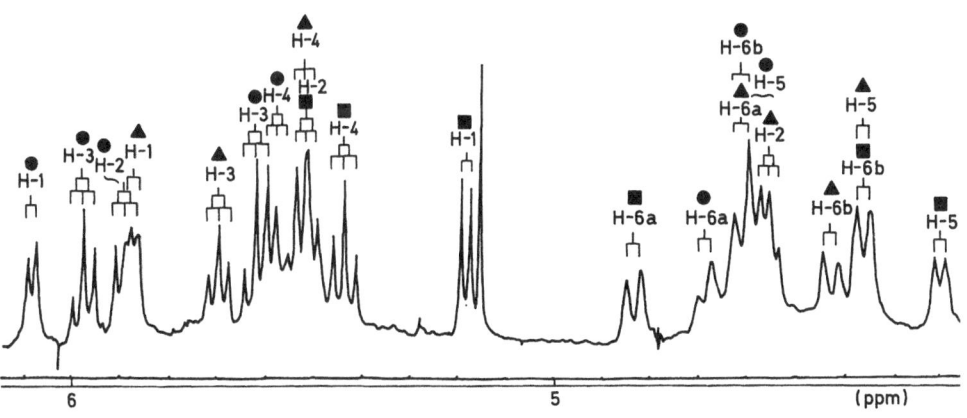

Fig. 7. ¹H-NMR spectrum of pertrifluoroacetate of tris-deacyl HBA (**18**). (**18**) was dissolved in trifluoroacetic anhydride and CDCl₃ (4:1) in an NMR tube. [From T. Goto *et al.*: Chemistry Lett. 883 (1981), with permission]

The ■-glucose was shown to be attached to the 2-position of ▲-glucose as follows. Tris-deacyl HBA (**18**) dissolved in trifluoroacetic anhydride and CDCl₃ (4:1) gave a solution of the pertrifluoroacetate of tris-deacyl HBA suitable for ¹H-NMR spectrometry (Fig. 7). The H-2 signal of ▲-glucose exhibited only a slight downfield shift, whereas large downfield shifts were observed for the other sugar signals except for H-5, indicating the attachment of the ■-glucose at the 2-OH of ▲-glucose. The above results confirmed the structure of tris-deacyl HBA (**18**).

Mild alkaline hydrolysis of HBA afforded *trans*-4-0-(6-0-(*trans*-3-0-(β-D-glucopyranosyl)caffeyl)-β-D-glucopyranosyl)caffeic   acid (mGCpGC, **19**) (*31*) and bis-deacyl HBA (**17**), which was further hydrolyzed to tris-deacyl HBA (**18**) and *trans*-3-0-(β-D-glucopyranosyl)caffeic acid (mGC, **20**). The acyl group present in tris-deacyl HBA (**18**) was deduced to be in the 6-position of ■-glucose by PRFT and NOE relief methods (*30*).

HBA (**9**) consists of bis-deacyl HBA (**17**) esterified with mGCpGC (**19**). The position of attachment of (**19**) on (**17**) was deduced by the

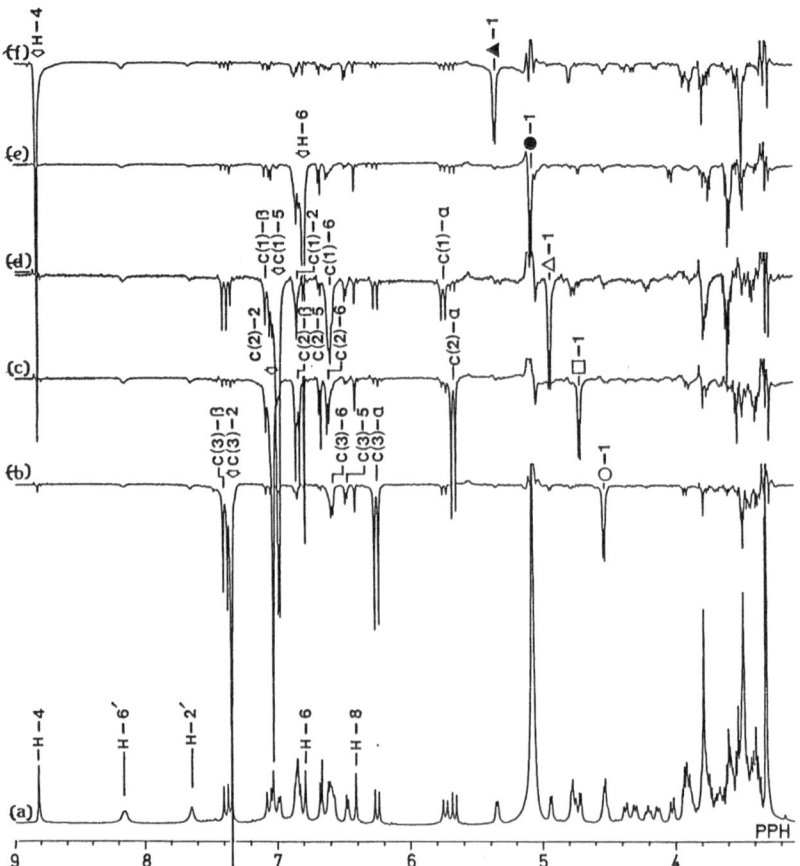

Fig. 8. ¹H-NMR difference spectra of HBA (**9**) by irradiation of the aromatic protons (8% CF₃COOD:CD₃OD at 5 °C, 500 MHz). (a) Normal spectrum; (b)–(f) NOE difference spectra by irradiation at the position indicated by the arrow. (Some long-distance NOE's are observed but their intensities are low so as not to interfere with the assignments.) [From T. Kondo *et al.*: Tetrahedron Lett. **28**, 2273 (1987)]

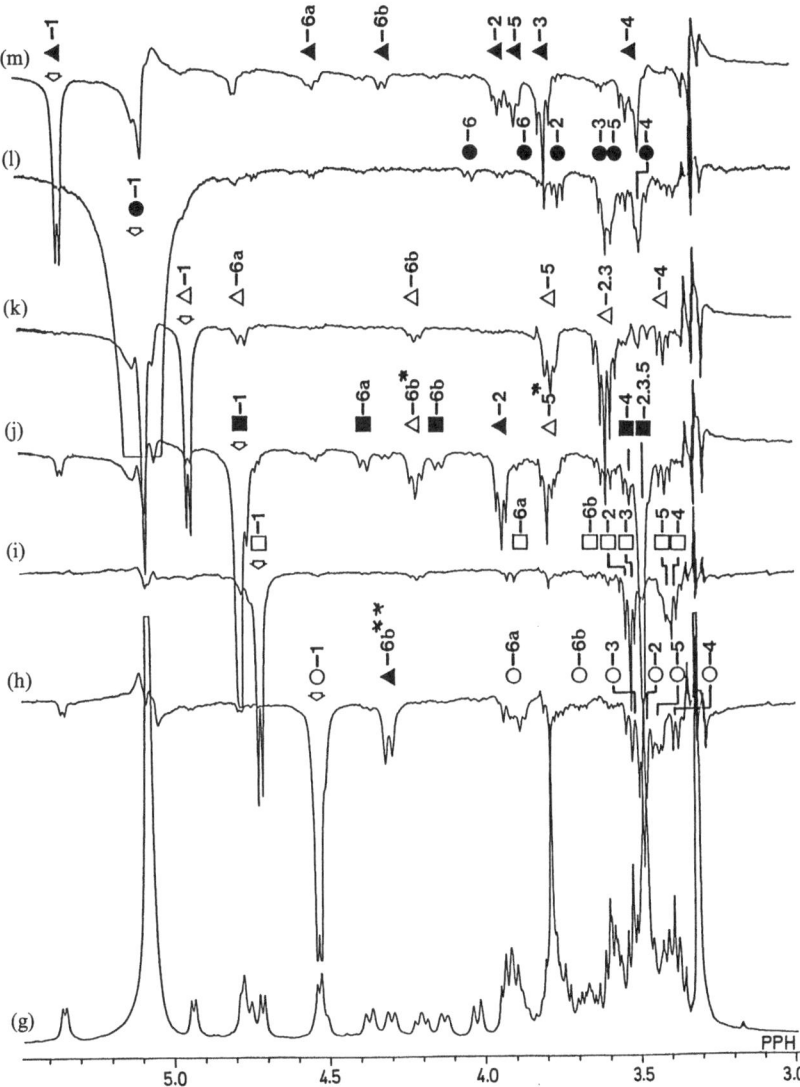

Fig. 9. High field region of ^{1}H-NOE difference spectra of HBA (9) by irradiation of the anomeric protons of the glucose moieties (8% CF$_3$COOD:CD$_3$OD at 5 °C, 500 MHz). (g) Normal spectrum; (h)–(m) NOE difference spectra by irradiation at the position indicated by the arrow. *Since Δ-6$_a$ is also irradiated, the NOE signals for Δ-6$_b$ and Δ-5 appear in this spectrum. **Since ▲-6$_a$ is also irradiated, the NOE signals for ▲-6$_b$ appear in this spectrum. [From T. KONDO *et al.*: Tetrahedron Lett. **28**, 2273 (1987)]

NOE relief method, a special application of negative NOE difference spectroscopy, as follows (68). The proton NOE of HBA was measured (500 MHz) in 8% TFA-d:CD$_3$OD at 5 °C (Fig. 8 and 9). Irradiation of the C(1)-5 proton signal produced the NOE difference spectrum (Fig. 8d) in which 2, 6, α, and β proton signals of the C(1) group as well as the anomeric proton (△-1) of a glucose moiety were relieved, indicating that the △-glucose is attached to the 4-OH of C(1). Similarly, irradiation of the C(2)-2 and the C(3)-2 protons disclosed that the □- and ○-glucose units are attached to the 3-OH of C(2) and C(3), respectively (Fig. 8b and 8c). That two glucose units, ▲ and ●, are attached, respectively, to the 3- and 5-positions of the peonidin nucleus was deduced from observation of NOE between ▲-1 and H-4, and between ●-1 and H-6 (Fig. 8f and 8e). Thus, all six anomeric protons of the glucose moieties could be correlated with peonidin and the three caffeic acid moieties.

Next, each anomeric proton on the six glucose moieties was irradiated as shown in Fig. 9h–m. Irradiation of the ▲-1 proton relieved all protons attached to the carbon atoms of ▲-glucose, whose two ▲-6 protons were at δ 4.30 and δ 4.52, indicating that ▲-CH$_2$OH is acylated. Since bis-deacyl HBA (17) has no acyl group on the ▲-glucose unit, the ▲-6 position in HBA (9) must be esterified with mGCpGC (19).

Platyconin (8)

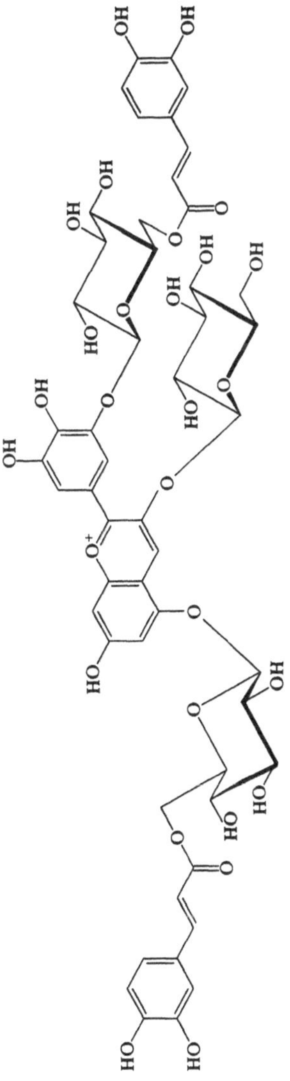

Gentiodelphin (7)

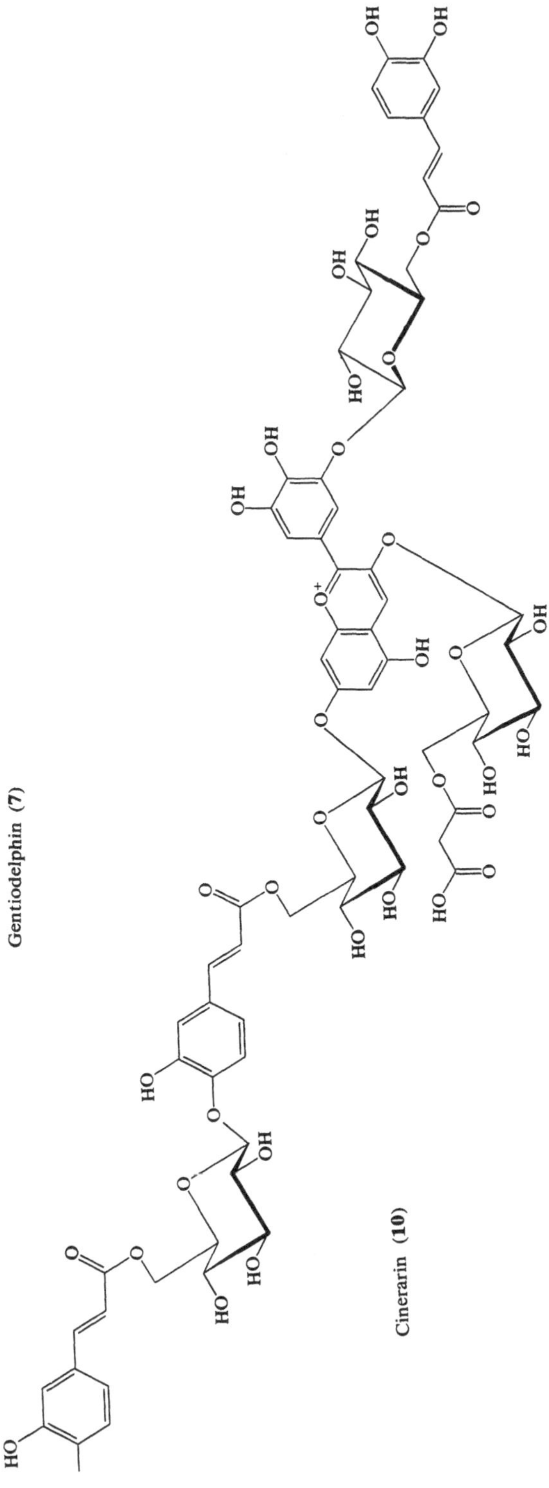

Cinerarin (10)

2.4 Other Anthocyanins Acylated with Two or More Aromatic Acids

In 1972 Saito et al. (86, 88) isolated platyconin from the Chinese bellflower *Platycodon grandiflorum* and proposed its structure to be delphinidin 3-dicaffeylrutinoside-5-glucoside. They also reported that bisdeacylplatyconin was identical with the deacylation product of violanin, delphinidin 3-rutinoside-5-glucoside (21). Gradient ODS HPLC showed, however, that they are not identical (34) and that bisdeacyl-platyconin was delphinidin 3-rutinoside-7-glucoside (22). Platyconin consists of bisdeacylplatyconin (22) and two molecules of *trans*-4-0-β-D-glucosylcaffeic acid (pGC, 23) as is evident from the FAB-MS and ^{1}H-NMR spectra, two pGC (23) being attached to the 6-CH$_2$OH of the two glucose moieties, which results in structure (8). On the other hand, Saito et al. (89) suggested by analysis of a high-resolution FAB-MS spectrum that platyconin was delphinidin 3-rutinoside 5-glucoside diacylated with glucosylcaffeic acid on the glucose moiety of rutinose.

Yoshitama (120, 121) isolated cinerarin from garden cineraria *Cenecio cruentus* and suggested that its structure was dicaffeyldelphinidin 3,7,3'-triglucoside, but it is actually malonyltricaffeyldelphinidin tetraglucoside (10). Gentiodelphin (7) (33) was isolated from the blue petals of *Gentiana makinoi*, and zebrinin (24) from violet leaves of *Zebrina pendula* (57). All of the cinnamic acid derivatives found in the acylated anthocyanins have a *trans* double bond; no *cis* isomer has been found in spite of a previous report (14) to the contrary.

2.5 Anthocyanins Acylated with Aliphatic Dicarboxylic Acids

Willstätter and Bolton (114) reported that salvianin isolated from *Salvia coccinea* consists of pelargonidin diglucoside acylated with malonic acid. Karrer and Widmer (66, 67) isolated a substance monardaein from *Monarda didyma* which they considered to be identical with salvianin and composed of pelargonidin diglycoside acylated with two molecules of malonic acid and one molecule of p-coumaric acid. Contrary to the above results Harborne (41) could not find malonic acid in any of these anthocyanins. Indeed the structure proposal for monardaein by Birkofer et al. (13) contained no malonic acid. Subsequently, Boom and Geissman (15), identified malonic acid in *Mimulus luteus* anthocyanin and Cornuz et al. (25) unambiguously characterized an anthocyanin isolated from the red Iceland poppy, *Papaver nudicaule*, as pelargonidin 3-malonylsophoroside (25). Cyanidin 3-malonylglucoside (14) was also isolated from *Cichorium intybus* (16).

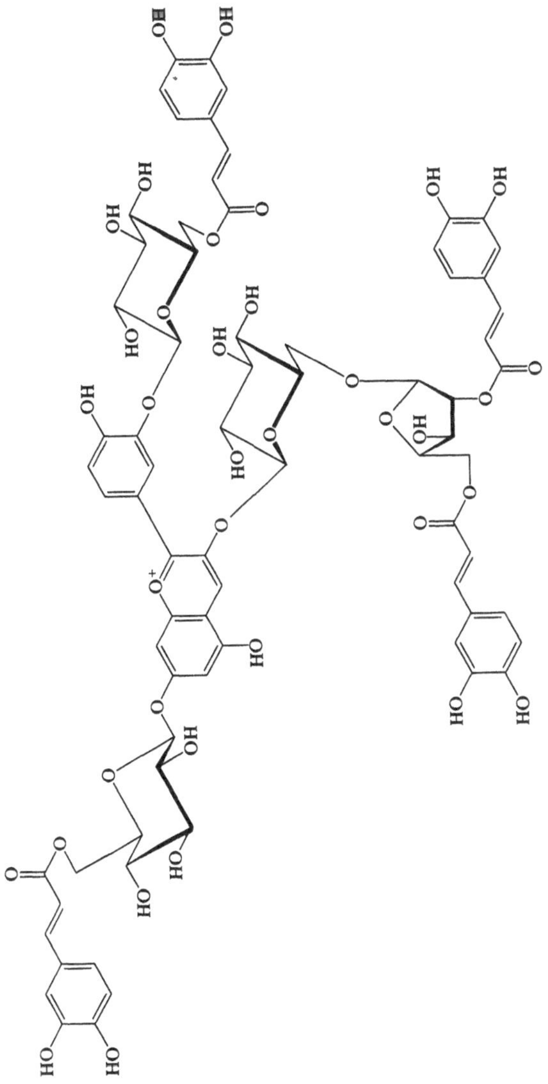

Zebrinin (24)

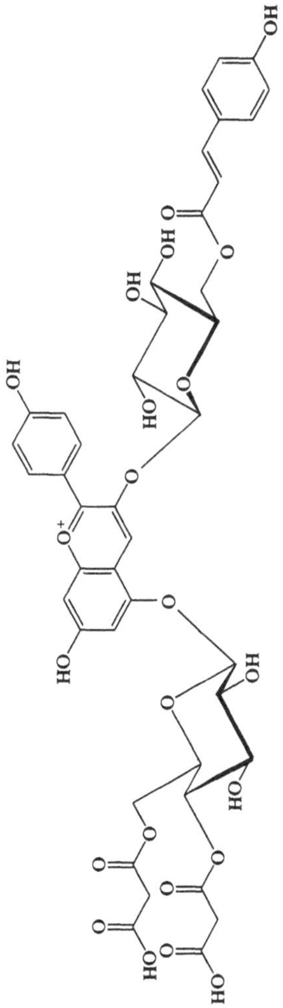

Monardaein (27)

A metalloanthocyanin, commelinin (26), isolated from *Commelina communis* by HAYASHI contains an anthocyanin, awobanin (12) (71, 72, 73), whose structure was shown to be delphinidin 3-(p-coumaryl)glucoside-5-glucoside (35, 74, 98). Later, however, awobanin (12) was found to be an artifact produced during extraction with acid and the real anthocyanin is malonylawobanin (11) (35). Similarly, cyanin

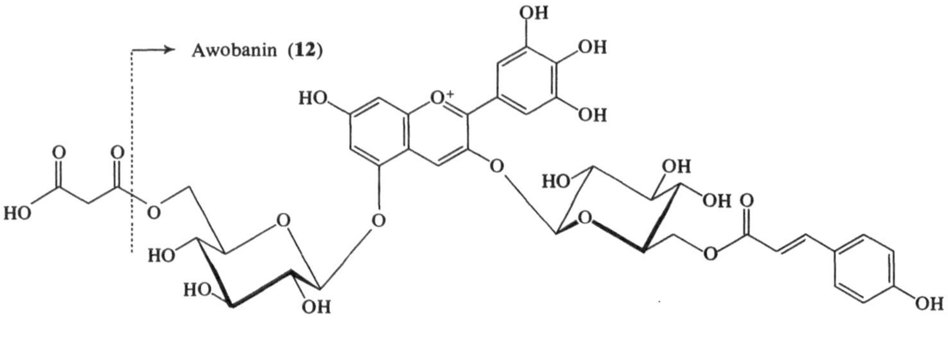

Malonylawobanin (11)

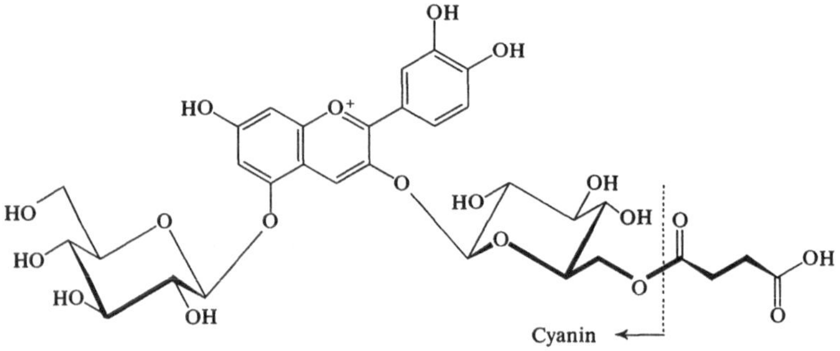

Succinylcyanin (13)

(2) isolated from cornflower, *Centaurea cyanus*, by WILLSTÄTTER and EVEREST (*115*) was found to be an artifact. The intact anthocyanin (13) incorporates succinic acid and was named succinylcyanin (*105*) or centaurocyanin (*94, 103*). Cinerarin (10) also contains malonic acid (*36*).

The above results suggest that such malonylated anthocyanins might well have been overlooked in earlier studies because of the ease of hydrolysis with methanolic hydrochloric acid used for extraction. Indeed monardaein (27) obtained by mild extraction with aqueous tri-fluoroacetic acid contained two molecules of malonic acid (*69*). Anthocyanins isolated from the blue flowers of *Clitoria terenatea*, which were named ternatin A–F, consist of delphinidin 3,3′,5′-triglucoside, *p*-coumaric acid and malonic acid (*85*).

HARBORNE (*44*) re-investigated natural anthocyanins by electro-phoresis, because malonated anthocyanins exist as zwitterions at around pH 4.4. He surveyed flowers of 81 species belonging to 27 plant families and found that zwitterionic anthocyanins occur in half of the samples in particular, almost all species in the Compositae and the Labiatae gave a positive test. Thus, zwitterionic anthocyanins are widespread in nature.

3. Stabilization Effect of Inorganic Salts

GOTO *et al.* (*27*) found that anthocyanin anhydrobases as well as flavylium ions are strongly stabilized and do not form colorless pseudobases by hydration when dissolved in concentrated aqueous solutions of some neutral salts such as magnesium chloride and sodium chloride. For example, in 4M $MgCl_2$ solution both the flavylium and the anhydrobase of 7,4'-dihydroxyanthocyanidin exist as such and do not tend to decompose (Fig. 10), while they decompose rapidly to form colorless products when dissolved in water.

Genuine anthocyanins, i.e. those whose structure is that of the compouds present in the plants themselves and are mostly in the form of an anhydrobase (*102*), can be extracted from plants quantitatively

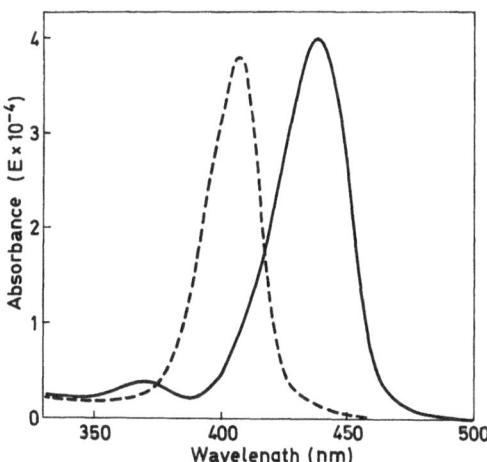

Fig. 10. Visible spectra of 7,4'-dihydroxyanthocyanidin (1.83×10^{-5} mole/liter). Both solutions are stable for a few hours. --- Flavylium ion (in 4 M $MgCl_2$ containing 5×10^{-4} M HCl); — Anhydrobase (the above solution was neutralized by addition of 0.05 ml conc. ammonia). [From T. GOTO, T. HOSHINO, and M. OHBA: Agric. Biol. Chem. **40**, 1593 (1976), with permission]

with salt solutions. For example, the anthocyanins in pansy petals of different colors can be extracted with 4 M $MgCl_2$ solutions; their UV-VIS spectra measured in 10 mm cells (ca 10^{-5} M solution) are reproduced in Fig. 11. At λ max of the visible absorption the ratio of intensities exhibited by the anhydrobase and the corresponding flavylium ion is almost constant in each case, indicating that the anhydrobase is extracted with little, if any, decomposition. The colors of the extracts are identical with those of petals. This method made it possible to obtain UV-VIS spectra of the anhydrobases in dilute aqueous solutions.

Fig. 12 shows the stability of commelinin (26) in salt solutions. Commelinin is completely stable in concentrated NaCl solution, while in water it decomposes rapidly. The linear relation between the concen-

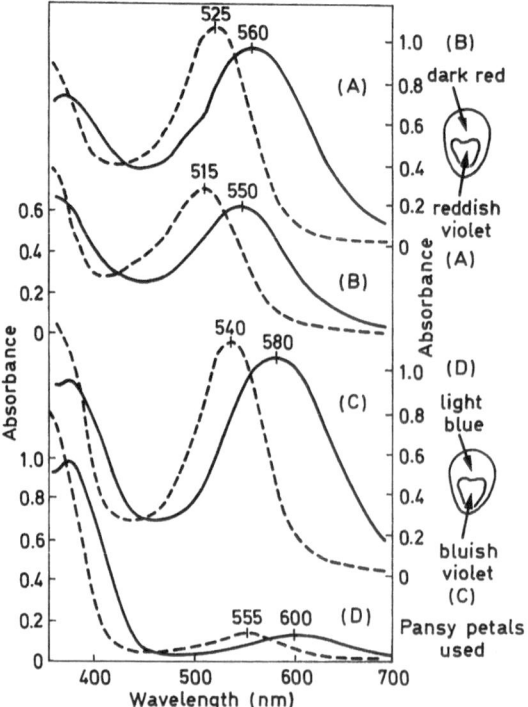

Fig. 11. Visible spectra of genuine anthocyanins of pansy petals in 4 M $MgCl_2$ solution. A disc (12 mm diameter) of petal was extracted with 4 M $MgCl_2$ a few times and the combined extracts were diluted with 4 M $MgCl_2$ to 10 ml, whose visible spectrum was measured in a cuvette of path length 10 mm. — in 4 M $MgCl_2$; --- in 4 M $MgCl_2$ (3 ml) containing 0.05 ml of conc. HCl. [From T. GOTO, T. HOSHINO and M. OHBA: Agric. Biol. Chem., **40**, 1593 (1976), with permission]

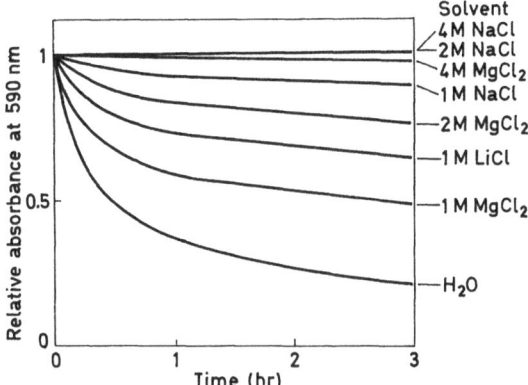

Fig. 12. Stability of commelinin in aqueous salt solutions (concentration: 31.3 mg/liter; path length 10 mm; 20 °C). [From T. GOTO, T. HOSHINO and M. OHBA: Agric. Biol. Chem. **40**, 1593 (1976), with permission]

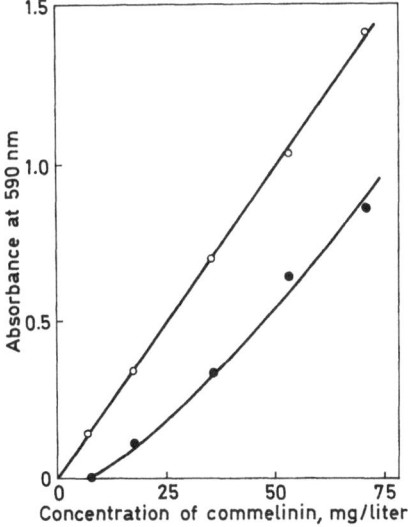

Fig. 13. Plots of the absorbance at 590 nm vs. concentration of commelinin. o—o in 4 M $MgCl_2$ (0.5 hr after dissolution at 20 °C); •—• in water. [T. GOTO, T. HOSHINO, and M. OHBA: Agric. Biol. Chem. **40**, 1593 (1976), with permission]

tration and absorbance in the salt solution indicates that no decomposition occurs on dilution (Fig. 13).

Stabilization in NaCl solution may be due to promotion of self-association of anthocyanins, whereas $MgCl_2$ may reduce the concentration of free water by hydration of magnesium ions (*51*).

4. Self-Association

ASEN *et al.* (*6*) observed that at pH 3.16 the absorbance of cyanidin 3,5-diglucoside (**2**) at λ max increased 300 times when the concentration was increased from 10^{-4} M to 10^{-2} M (100 times), suggesting self-association of the anthocyanin flavylium ions. Malvidin 3-glucoside shows similar behavior at pH 3.5 (*107*). ASEN *et al.* (*7*) also observed that a change in the shape of the visible absorption spectrum occurred with dilution of delphinidin 3-di(*p*-hydroxybenzoyl)glucosylglucoside solution at pH 6.6 (anhydrobase) from 8×10^{-3} M to 8×10^{-5} M; the absorbance ratio $A_{624\,nm}/A_{540\,nm}$ changed from 0.58 to 1.10, resulting in a much bluer solution. Raising the temperature produced a similar change. They suggested that dilution or heating caused the associated form to change to the unassociated form of the anthocyanin.

On studying the co-pigmentation effect of rutin (**28**) (a flavone) on malvin (**6**) (see Chapter 5), SCHEFFELDT and HRAZDINA (*90*) observed that at higher concentrations of malvin at pH 3.20 the co-pigmentation effect became insignificant. They attributed this phenomenon to the self-association of malvin flavylium salts in high concentrations as suggested earlier (*6*); the self-associated anthocyanin cannot form complexes with the co-pigment and thus causes a decrease of the co-pigmentation effect.

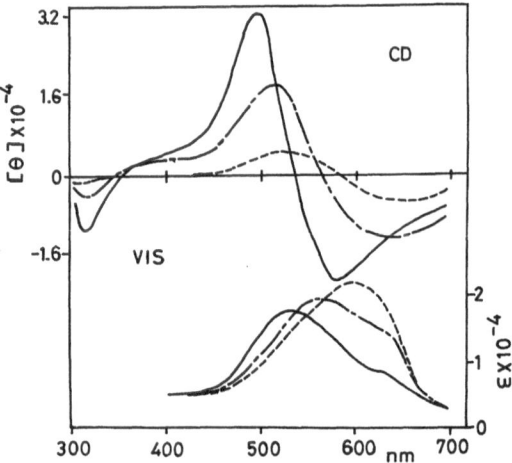

Fig. 14. Visible absorption and CD spectra of malvin (**6**) at pH 7.0 immediately after dissolving. --- 5×10^{-5} M; --- 5×10^{-4} M; — 5×10^{-3} M. [From T. HOSHINO *et al.*: Phytochem. **20**, 1971 (1981), with permission]

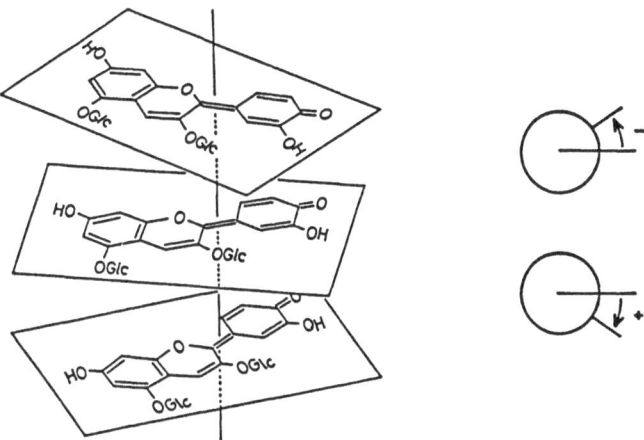

Fig. 15. A model of chiral stacking of anthocyanin molecules. [From T. GOTO *et al.*: Ann. New York Acad. Sci. **471**, 155 (1986), with permission]

More recently, HOSHINO *et al.* (*49, 53*) obtained firm evidence for self-association of anthocyanin anhydrobases in neutral solution by using circular dichroism as a sensitive probe for molecular association. Thus, the CD curves of peonin (**3**), delphin (**4**), hirsutin (**29**) and malvin (**6**) [for cyanin (**2**) and pelargonin (**1**), *vide infra*] show a large exciton-type splitting with first negative and second positive Cotton effects in the visible absorption band region when the anthocyanin chlorides are dissolved in a neutral buffer (pH 7.0) in a fairly high concentration (Fig. 14), indicating that two or more anthocyanidin chromophores must stack vertically in a left-handed screw axis to cause this type of split bands (Fig. 15) (*53*). The magnitudes of [θ] in each of the anthocyanins are strongly intensified when the concentration is increased, indicating enhanced formation of stacked molecules in the more concentrated solutions.

Shifting of the visible spectra towards shorter wavelengths due to formation of the chiral aggregates is also observed. For example, self-association of delphin anhydro base produces a remarkable hypso-chromic shift in the visible absorption as well as a large Cotton effect. The ^{1}H-NMR spectrum of malvin at pH 7.0 also indicated vertical stacking of the malvin molecules (*52*); the signals of the aromatic nucleus were all shifted toward higher field as the solute concentration increased, possibly as the result of diamagnetic anisotropy caused by ring currents in the neighboring aromatic molecules.

The driving force of this stacking is attributed to the hydrophobic interaction between the aromatic nuclei rather than to hydrogen bond-

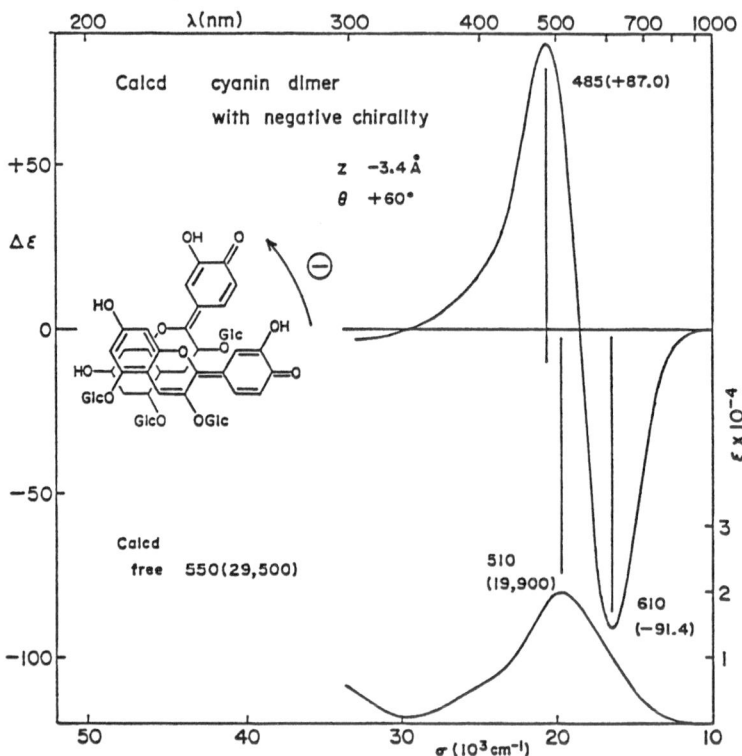

Fig. 16. Calculated CD and visible spectra of cyanin dimer, which are applicable to delphin dimer. [From T. Goto et al.: Ann. New York Acad. Sci. **471**, 155 (1986), with permission]

ing, because addition of urea or an organic solvent such as DMSO strongly destroys this stacking (51) and hirsutin (**29**), whose anhydrobase has no hydroxy group, is also capable of self-association. Whereas malvidin 3,5-diglucoside (**6**) and 5-glucoside show strong self-association, the [θ] values and the stability of malvidin 3-glucoside show little concentration dependence suggesting that malvidin 3-glucoside undergoes little self-aggregation (51).

Strong support for this type of stacking was obtained by nonempirical calculations. Fig. 16 shows calculated CD and visible absorption curves using exciton theory and SCF-CI-DV molecular orbital theory (38). The result is that, if the chromophores are stacked in a left-handed screw axis and the angle between two transition moments of the anthocyanin molecules is smaller than 90°, the absorption maximum shifts toward shorter wave length, that is, the substance exhibits a hypsochromic shift, and the CD curve exhibits a negative first Cotton and a positive second Cotton effect. The example shown in Fig. 16 has an angle of 60° between the two transition moments. The 40 nm hypsochromic shift is apparent by comparing the λ_{max} of the free and stacked molecules. The magnitude of the Cotton effect is in accord with the values observed for the anthocyanin aggregates discussed above. The concentration of anthocyanin in flower petals is around 10^{-2} M (7); this concentration is sufficient to produce this type of staking. Therefore, it is suggested that the stability and color variation of these anthocyanins in flower petals derive at least in part from the self-association.

Cyanin (2) showed unusual properties; the positive exciton type Cotton effect, whose sign was opposite to that of the case discussed above, increased gradually after the substance was dissolved in neutral buffer solution (49, 53). This aging lasted 60 min at which find blue floccules precipitated out. These, on being redissolved in water showed an exceptionally large exciton type Cotton effect, indicating very high stacking in a right-handed screw axis. This unusual phenomenon had its explanation in the fact that the cyanin floccules actually consisted of a metal chelate (39). Immediately after solution in neutral buffer cyanin indeed showed negative exciton-type Cotton effects of a magnitude similar to those of delphin (4) and malvin (6) (Fig. 17). The UV and CD spectra of the blue floccules of cyanin reported by HOSHINO

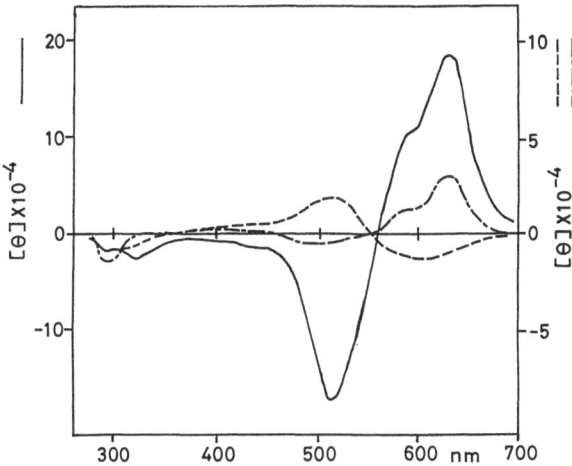

Fig. 17. CD spectra of cyanin anhydrobase (2) $(5 \times 10^{-4}$ M) in 0.1 M phosphate buffer at pH 7.0 (39). --- 2 min after dissolving without polyvalent metal ions; — Al^{3+} added; --- insoluble particles formed by aging without polyvalent metal ions (dispersed by ultrasonification)

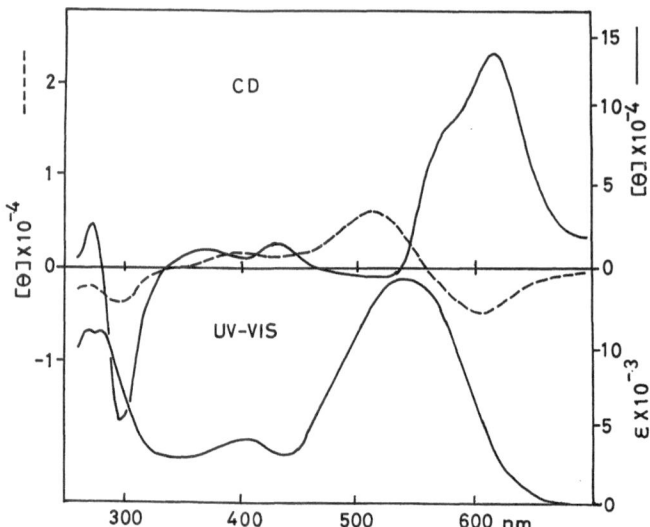

Fig. 18. Electronic and CD spectra of pelargonin (**1**) (7×10^{-4} M) in 0.05 M phosphate buffer at pH 7.27 (path length 1 mm) (*39*). CD: --- 2 min after dissolving; — insoluble particles formed by aging. UV-VIS: — 2 min after dissolving

et al. (*53*) were identical with those of the Al^{+++} complex ($\theta = 200,000$) of cyanin. Thus, the aging phenomenon of cyanin could be explained in terms of the slow formation of a metal chelate complex with trace metal ions supplied from the buffer solution and/or the glass vessel used.

Pelargonin (**1**) has no *o*-dihydroxy group and hence cannot form a metal complex, but curiously it shows an aging effect and a positive exciton-type Cotton effect similar to those of cyanin (**2**) (*53*). The CD curve, however, has an unsymmetrical shape; the first (positive) Cotton effect is large and the second (negative) Cotton effect is small (Fig. 18). Reexamination (*39*) disclosed that pelargonin (**1**) in a neutral buffer first showed a small negative exciton-type Cotton effect similar to that of other anthocyanins and then gradually formed a suspension of insoluble particles, which were responsible for the unsymmetrical positive exciton-type Cotton curve. The supernatant obtained by centrifugation showed no Cotton effect, whereas the CD curve was restored by dispersing the precipitates again in water ultrasonically. Since the solubility of pelargonin anhydrobase in water is very low, highly aggregated particles would have been formed.

In the complete absence of di- or trivalent metal ions, cyanin (**2**) also formed insoluble particles of a similar type which showed an unsymmetrical exciton-type CD curve with a first large positive and a second small negative sign (Fig. 17).

In conclusion, the self-association of anthocyanidin 3,5-diglucosides in neutral aqueous solution produces a negative exciton-type CD spectrum whose magnitude $[\theta]$ is around 1×10^4, indicating chiral stacking with the configuration of a left-handed screw.

5. Co-Pigmentation (*19, 42, 56, 80, 92*)

In 1931 ROBINSON found that flavones and tannins exert bathochromic and stabilization effects on anthocyanins and called this phenomenon co-pigmentation (*83*). He also pointed out (*84*) that insoluble polymers such as polysaccharides and polypeptides have similar effects. Co-pigmentation effects have been studied extensively, mostly between an anthocyanidin 3,5-diglycoside and a flavone: for example, malvidin glucosides and rutin (**28**) (*90*), delphanin (**30**) and C-glycosyl-flavones in Prof. Blaauw Iris (*9*), cyanidin glycosides and quercetin (**31**) glycosides in red wing azalea (*5*), cyanidin diglycosides and quercetin glycoside in rose "Better Times" (*4*), ophionin (**32**) and kaempferol (**33**) diglucoside in blue seed coats of *Ophiopogon jaburan* (*59, 63*), and malvin anhydrobase and C-glycosylflavone in *Fuchsia hybrida* (*116, 117*). In the case of *Hydrangea macrophylla*, it was suggested (*100*) that the blue color of the sepals is the result of co-pigmentation between delphinidin 3-monoglucoside and 3-*p*-coumarylquinic acid or 3-caffeylquinic acid; 5-caffeylquinic acid (chlorogenic acid) has no effect on the bluing.

Co-pigmentation (*5, 6, 116*) is affected by type and concentration of anthocyanin, type and concentration of flavone, and pH and temperature of the solution (*4, 113*). Those anthocyanins containing an aro-

Rutin   (**28**): $R_1 = OH$ $R_2 = $ rutinose
Quercetin (**31**): $R_1 = OH$, $R_2 = H$
Kaempferol (**33**): $R_1 = R_2 = H$

matic acyl group form a much stabler co-pigmentation complex with
C-glucosylflavones than unacylated anthocyanins (50), but anthocyanins having two or more aromatic acyl groups are stabilized by the intramolecular sandwich type stacking discussed in Chapter 6 rather than
by co-pigmentation. Co-pigmentation occurs with the flavylium ion
as well as the anhydro base (6) and only in solution non-stoichiometri-
cally, giving bluish-violet colors at the most and never producing a
pure blue color (4, 5, 9, 24, 90, 113).

Co-pigmentation causes not only a bathochromic shift in the visible
λ_{max} of anthocyanins but also results in a large increase in absorptivity
(stabilization). Figs. 19 and 20 show a linear relation between $\Delta A/mm$
at λ_{max} (nm) and $\Delta\lambda_{max}$ (nm) of flavylium ions. The bathochromic effect
of flavonoid co-pigments is more prominent with malvin (6) than with
cyanin (2).

For a quantitative description of the co-pigmentation effect, Hos-
HINO et al. (50) introduced a co-pigmentation constant Kc and an affini-

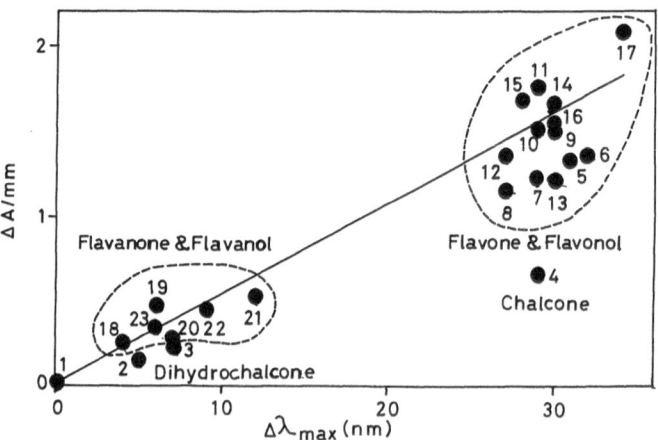

Fig. 19. Co-pigmentation of flavonoid compounds and phloroglucinol derivatives
$(4 \times 10^{-3}$ M) with malvin (6) $(1 \times 10^{-3}$ M) at pH 3.2 (path length 1 mm). ΔA: difference
between the absorbance at λ_{max} of visible absorption of malvin with and without a co-
pigment. $\Delta\lambda_{max}$: difference between the λ_{max} of visible absorption of malvin with and
without a co-pigment. Co-pigments: 1 phloroglucinol, 2 phloroacetophenone, 3 phlorid-
zin, 4 poncirin (chalcone), 5 apigenin 7-glc., 6 apigenin 7-neohesperidoside, 7 kaempferol
7-neohesperidoside, 8 kaempferol 3-robinobioside-7-rhamnoside, 9 quercetin 3-glc.,
10 quercetin 3-rham., 11 quercetin 3-rutinoside, 12 quercetin 7-neohesperidoside, 13 isor-
hamnetin 3-glc., 14 tamarixetin 7-rutinoside, 15 quercetin, 16 fisetin, 17 myricetin, 18 pin-
ocembrin 7-neohesperidoside, 19 naringenin 7-neohesperidoside, 20 isosakuranetin 7-
neohesperidoside, 21 eridicyol 7-neohesperidoside, 22 taxifolin, 23 (+)-catechin. [Data
taken from L.-J. CHEN and G. HRAZDINA: Phytochem. 20, 297–303 (1981), with permis-
sion]

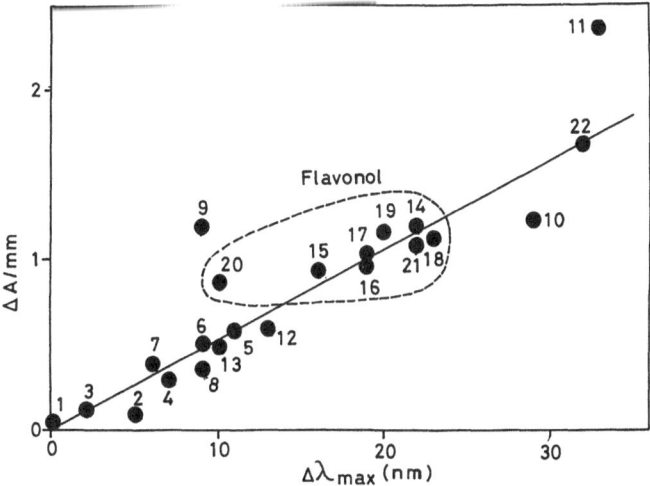

Fig. 20. Co-pigmentation of cyanin (2) (2×10^{-3} M) with a co-pigment (6×10^{-3} M) at pH 3.32 (path length 1 mm). ΔA and Δλ: see the legend of Figure 19. Co-pigments: *1* glycine, *2* caffeine, *3* protocatechuic acid, *4* caffeic acid, *5* sinapic acid, *6* phloridzin, *7* (+)-catechin, *8* apigenin 7-glc., *9* 8-C-glucosylapigenin, *10* 6-C-glucosylapigenin,*11* 6-C-glucosylgenkwanin, *12* hesperidin, *13* naringin, *14* kaempherol 3-glc., *15* kaempherol 3-robinobioside-7-rhamnoside, *16* quercetin 3-glc., *17* quercetin 3-rham., *18* quercetin 3-gal., *19* quercetin 3-rutinoside, *20* quercetin 7-glc., *21* 7-O-methylquercetin 3-rham., *22* aureusidin (aurone). [Data taken from S. Asen, R.N. Stewart and K.H. Norris: Phytochem. **11**, 1139–1144 (1972), with permission]

Fig. 21. Hydrophobic stacking type co-pigmentation model. [From T. Goto, T. Hoshino, and S. Takase: 20th Symposium on the Chemistry of Natural Products. Symposium papers, 59–66 (1976)]

ty index 1/Kc. The co-pigmentation constant (Kc) is defined as the molar ratio of co-pigment to anthocyanin which yields half-maximal absorbance at a given concentration of the anthocyanin. Comparison of Kc values is only possible at the same molar concentration of anthocyanins; for example, the Kc values of anthocyanin (5×10^{-4} M) and

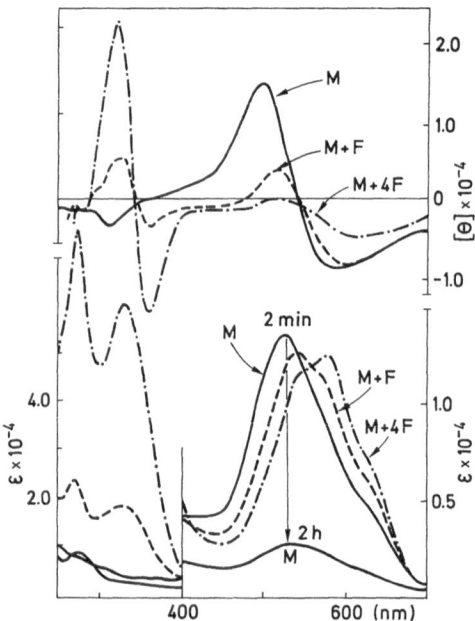

Malvin (A) (6)   Flavocommelin (F) (34)

Self-association   Co-pigmentation

Fig. 22. Schematic representation of self-association of malvin (6) and co-pigmentation between malvin (6) and flavocommelin (34)

Fig. 23. CD and electronic spectra of malvin and malvin-flavocommelin complexes in 0.01 M phosphate buffer at pH 6.0, 2 min after dissolving (concentration of malvin: 2×10^{-3} M). M: malvin (6); F: flavocommelin (34). — M; --- $M:F=1:1$; --- $M:F=1:4$. [From T. Goto et al.: Ann. New York Acad. Sci. 471, 155 (1986), with permission]

flavocommelin at pH 6.0 are: tubouchinin (p-coumarylmalvin) 1.6, malvin 8.0, and delphin 12.0.

The co-pigmentation effect had long been considered to come from hydrogen bonding between the host compound and the co-pigment used (*12, 24, 90*) producing horizontal stacking. WILLIAMS and HRAZ-DINA (*113*) assumed that hydrogen bonds form between the keto group of anthocyanins and the hydroxyl group of the flavonol glycosides to give horizontal association of the pigments and that the degree of complex formation depends mainly on the number of free aromatic hydroxyl groups in the flavonoid (*24*). GOTO et al. (*28, 29*) proposed that co-pigmentation in aqueous solution arises from hydrophobic stacking between the aromatic nuclei of anthocyanin and flavone (vertical stacking) (Fig. 21). This hydrophobic stacking could be stabilized further by the hydrophilic sugar moieties covering them. SWEENY et al. (*95*) also supported this mechanism.

An example is shown in Fig. 22 (*38*). Malvin anhydro base (**6**) in aqueous solution (5×10^{-4} M) at pH 6.0 is almost decolorized after

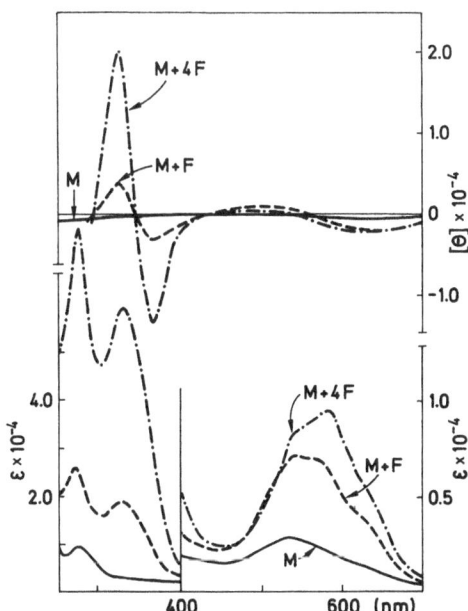

Fig. 24. CD and electronic spectra of malvin and malvin-flavocommelin complexes in 0.01 M phosphate buffer at pH 6.0, 2 hours after dissolving (abbreviations and symbols same as in Figure 23). [From T. GOTO et al.: Ann. New York Acad. Sci. **471**, 155 (1986), with permission]

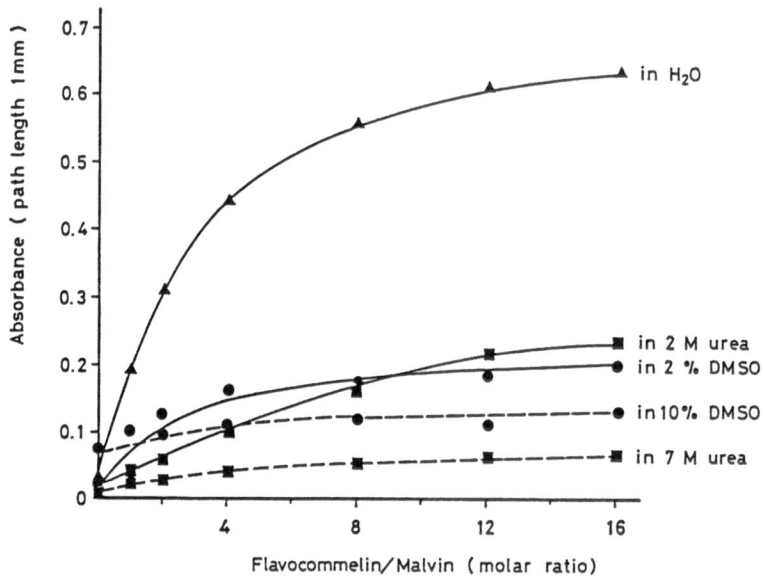

Fig. 25. Effect of urea and dimethyl sulfoxide on stability of copigmented complex in 0.01 M phosphate buffer at pH 6.0

2 hours, but it is stabilized very much by addition of flavocommelin (**34**) and the color of the solution is also changed from reddish violet to bluish violet (*50*), a 50 nm bathochromic shift is observed. A charge transfer interaction would also be operating (*79*). This phenomenon is concentration dependent and at a higher concentrations of malvin the co-pigmentation effect of flavocommelin is strongly enhanced. On the other hand, the exciton type Cotton effect exhibited by malvin in aqueous solution disappears on addition of flavocommelin (Fig. 23 and 24), indicating that the self-association of malvin (**6**) is destroyed by flavocommelin (**34**) to form a new intermolecular malvin-flavocommelin complex (co-pigmentation). In turn, this co-pigmentation is inhibited by addition of dimethyl sulfoxide or urea (Fig. 25), possibly by weakening the hydrophobic interaction between malvin and flavocommelin (*51*).

Incidentally, flavocommelin (**34**) itself also displays chiral stacking as indicated by its exciton type CD around 340 nm whose intensity is concentration and temperature dependent (Goto *et al.* unpublished). Thus, chiral stacking may be quite common among glycosides of naturally occurring aromatic compounds.

References, pp. 153–158

6. Intramolecular Sandwich Type Stacking

SAITO *et al.* (*86*) found that platyconin (**8**) is unusually stable in neutral or weakly acidic aqueous solutions in which common anthocyanidin mono- and di-glycosides are unstable and become almost colorless. Cinerarin (**10**) was also found by YOSHITAMA (*120, 121*) to be very stable in such aqueous solutions. It was suggested that anthocyanins containing two or more aromatic acyl groups, such as hydroxylated cinnamyl residues, are stable in neutral or weakly acidic solutions possibly as the result of hydrogen bonding between phenolic hydroxyl groups in anthocyanidin and aromatic acids (*120*). BROUILLARD (*17, 19*) and GOTO *et al.* (*33, 34*) considered that diacylated anthocyanins must be stabilized by a sandwich type stacking (Fig. 26) caused by hydrophobic interaction between the anthocyanidin ring and the two aromatic acyl groups as suggested in the cases of co-pigmentation (*28, 29*) and self-association (*49, 51, 52, 53*).

Some evidence has been obtained from NMR spectrometry (*38*). Careful nuclear Overhauser effect (NOE) measurements of heavenly blue anthocyanin (HBA) (**9**) showed several long-distant NOE's (Fig. 27), indicating a folded conformation for HBA. In comparison with those of methyl glucosylcaffeates, the proton signals on the two caffeic acid residues which are close to the anthocyanidin nucleus of HBA exhibited large upfield shifts, whereas the signals of the third

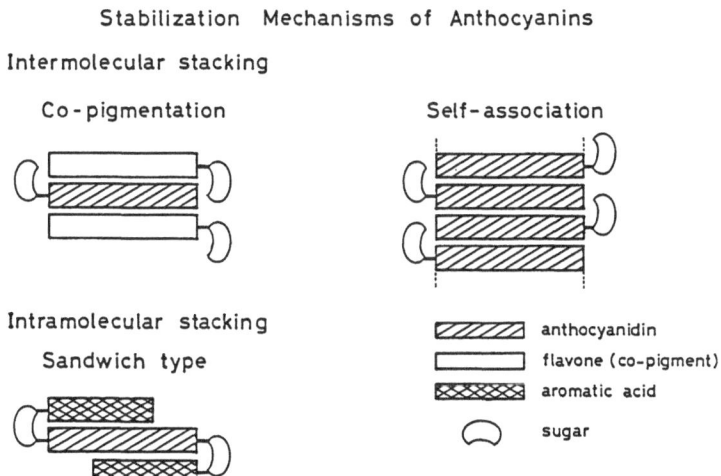

Fig. 26. Three stacking mechanisms of anthocyanins. [From T. GOTO *et al.*: Ann. New York Acad. Sci. **471**, 155 (1986), with permission]

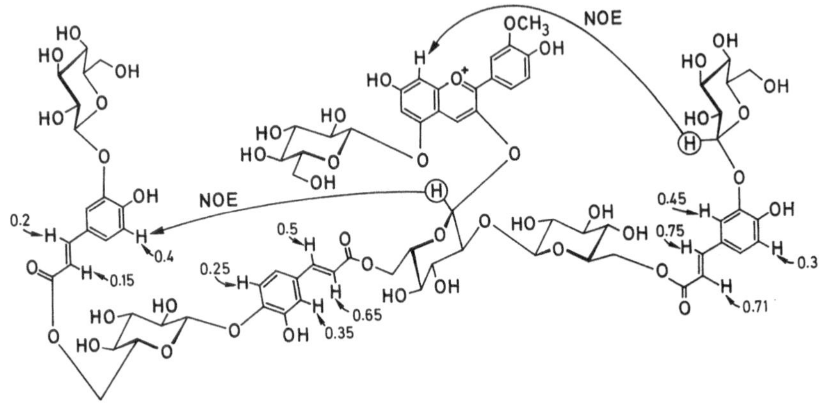

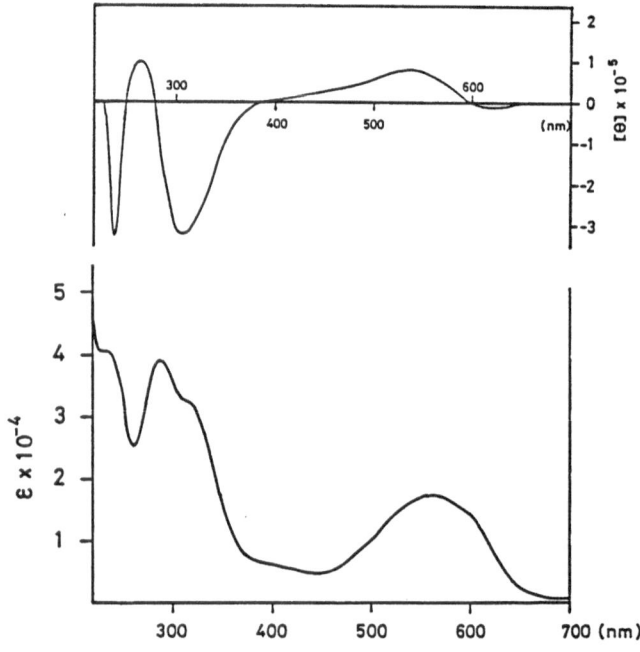

Fig. 27. The up-field shifts (ppm) and long-distance NOE observed in the ¹H-NMR spectrum of heavenly blue anthocyanin (**9**). Numbers indicate the chemical shift differences
(ppm) between the ¹H-NMR signal (at upper field) of heavenly blue anthocyanin and
the corresponding signal of methyl glucosylcaffeate (solvent: CD₃OD + DCl). [From T.
Goto *et al.*: Ann. New York Acad. Sci. **471**, 155 (1986), with permission]

Fig. 28. CD (upper) and electronic (lower) spectra of heavenly blue anthocyanin (**9**)
(5 × 10⁻³ M) in water at pH 7.0. [From T. Goto *et al.*: Ann. New York Acad. Sci. **471**,
155 (1986), with permission]

caffeic acid residue which is distant from the anthocyanidin showed only small shifts, indicating that two of the aromatic acyl groups are stacked parallel with the anthocyanidin nucleus. The CD curve showed a strong Cotton effect near the absorption of the acyl groups (ca. 340 nm), whereas no exciton type splitting was observed in the visible absorption of the anthocyanidin part (Fig. 28).

Fig. 29 shows an intramolecular sandwich type stacking model of HBA. Such stacking prevents the HBA molecule from reacting with water to form the pseudobase, thus stabilizing the flavylium ion as well as the anhydro base. The blue color of the petals is caused by an increase in the pH of the cell sap to 7.5 (8, 62), an unusually high pH. Fig. 30 shows that the acyl groups in HBA strongly stabilize the anthocyanidin chromophore.

Fig. 29. Molecular models of heavenly blue anthocyanin (9). (a) Extended form; (b) sandwich-type stacking form. [From T. GOTO et al.: Ann. New York Acad. Sci. **471**, 155 (1986), with permission]

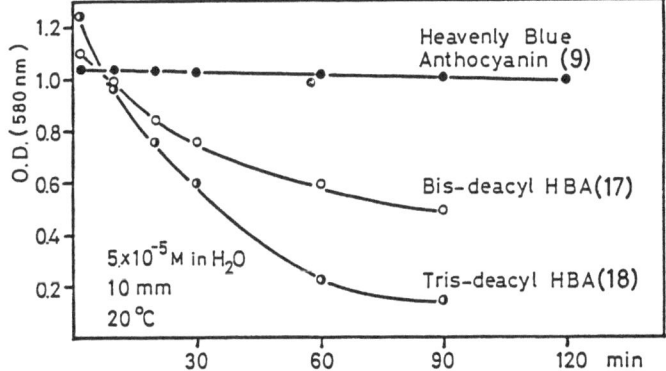

Fig. 30. Stability of heavenly blue anthocyanin and its deacylated derivatives in water. [From T. GOTO et al.: Ann. New York Acad. Sci. **471**, 155 (1986), with permission]

Gentiodelphin (**7**) (*33*), platyconin (**8**) (*34*), cinerarin (**10**) (*36*) and zebrinin (**24**) (*57*) are also stable in weakly acidic or neutral aqueous solutions. For all of them sandwich type conformations similar to that of HBA are possible. The following acylated anthocyanins, whose structures are not completely elucidated, have also been found to be stable in water: *Lobelia* anthocyanin from *Lobelia erinus* (*118*), ternatin A to F isolated from the blue flowers of *Clitoria ternatea* (*85*), and a blue and a purple anthocyanin isolated from the flowers of *Tradescantia reflexa* (*119*).

7. Metalloanthocyanins

Metalloanthocyanins consist not only of anthocyanin and metal ion(s) but also contain a co-pigment such as a flavone. Until now only commelinin and protocyanin have been characterized as metalloanthocyanins. The blue color of *Hydrangia* has been considered to stem from a metalloanthocyanin, but this opinion is contradicted in a recent report (*100*).

7.1 Commelinin

WILLSTÄTTER in 1913 attributed the variety of flower colors to pH variation in flower cell sap (*115*). Although a blue color can be produced in alkaline pH, it fades very rapidly. SHIBATA *et al.* (*91*) observed in 1919 that reduction of the flavone with magnesium and hydrogen chloride in ethanol afforded a blue anthocyanin that contained magnesium and proposed the so-called metal complex theory according to which the blue color is produced by complexation of anthocyanins with metal ions such as magnesium ion. EVEREST, one of the collaborators of WILLSTÄTTER, strongly opposed this theory (*26*). Since then the theory suffered a long period of neglect, but was revitalized in 1957 by the work of HAYASHI *et al.* (*46, 76, 77*) who isolated from deep blue petals of *Commelina communis* a blue-colored anthocyanin, commelinin (**26**) in crystalline form. Commelinin consists of two molecules each of an anthocyanin, awobanin (A) (**12**), and a flavone, flavocommelin (F) (**34**), and one atom each of magnesium and potassium. Its color and stability were explained in terms of a co-ordinated complex of magnesium and four molecules of the flavonoids (A and F) (Fig. 31) (*48, 101*). Such metal complexes of anthocyanin were named

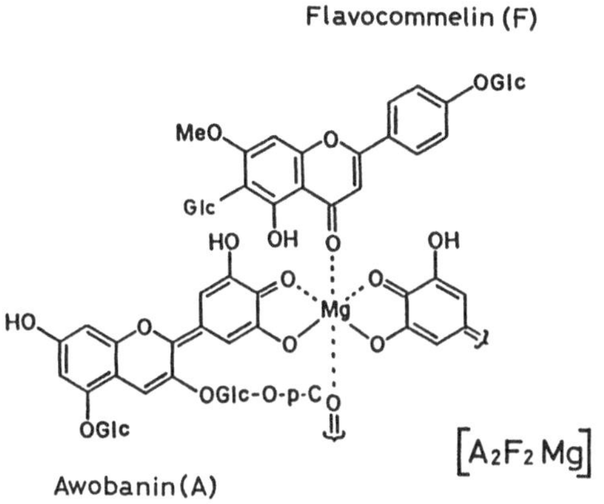

Flavocommelin (F)

$$\left[A_2 F_2 Mg\right]$$

Awobanin (A)

Fig. 31. Structure of commelinin proposed by HAYASHI and TAKEDA. [From K. HAYASHI and K. TAKEDA: Proc. Japan Acad. **46**, 535 (1970)]

metalloanthocyanins (*76*). BAYER disagreed with this explanation, however, because in general magnesium ion does not form stable chelates with anthocyanins (*12*).

GOTO *et al.* assumed that in commelinin awobanin (A) and flavocommelin (F) are stacked in parallel by hydrophobic interactions between the aromatic nuclei (Fig. 21) and that magnesium ion plays no important role in the stacking (*29*). However, TAKEDA and HAYASHI (*96, 97, 99*) showed that magnesium ion is indeed necessary, but that

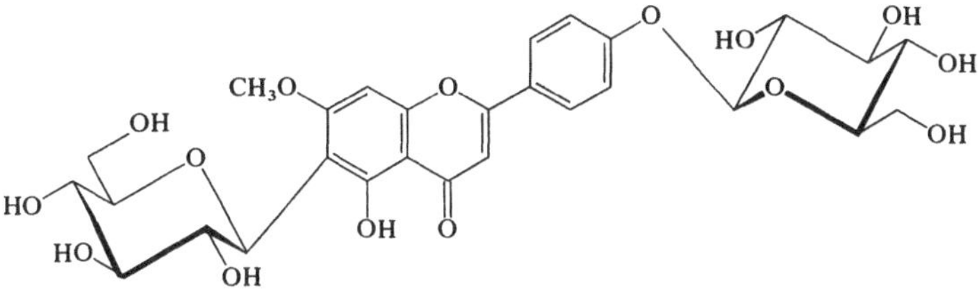

Flavocommelin (**34**)

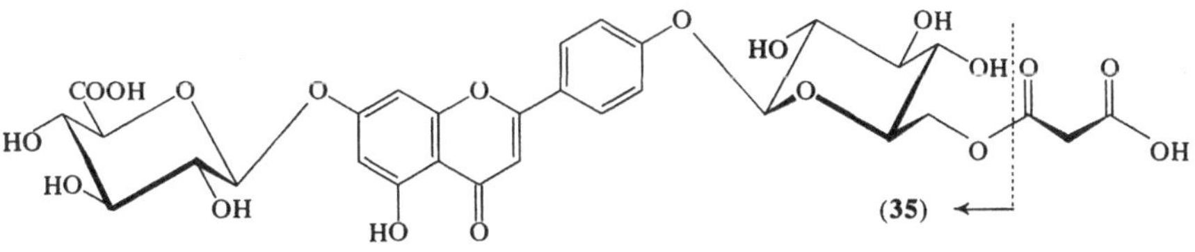

Apigenin 4-O-(6-O-malonyl-β-D-glucoside)-7-O-β-D-glucuronide (**36**)

zinc, cadmium, cobalt or manganese ions can also form complexes very similar to commelinin. On electrophoresis commelinin moves toward the positive pole (*46*, *81*) and hence it must have a negative charge(s). This observation long remained mysterious, since neither awobanin (A) (**12**) nor flavocommelin (F) (**34**) has a negative charge, nor does magnesium ion. However, this problem was solved by the discovery that the true anthocyanin present in the flower of *Commelina*

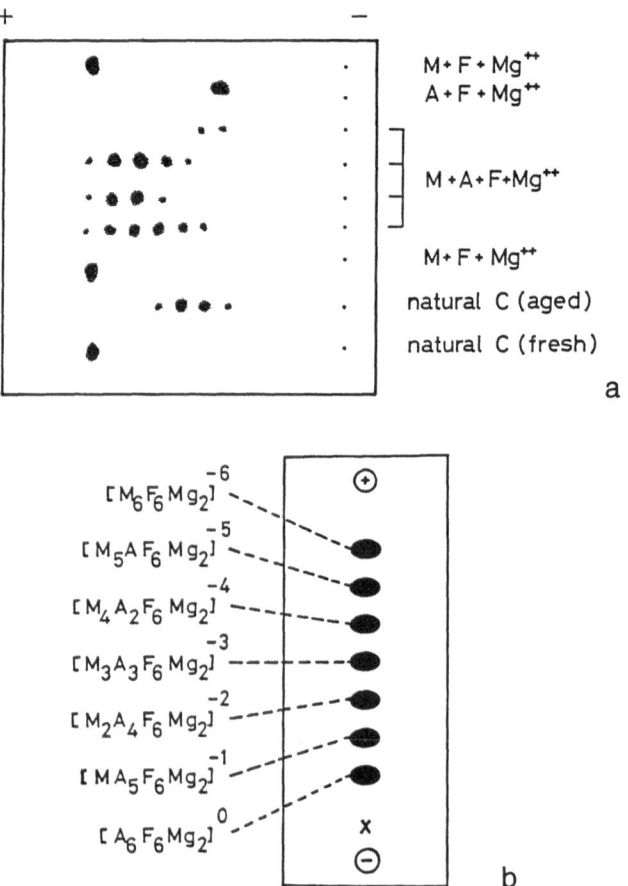

Fig. 32. (a) Electrophoresis of natural and synthetic commelinin and demalonylcommelinins (cellulose acetate film; 0.04 M acetate buffer at pH 5.7 for 2 hours at 300 V). *A*: awobanin (**12**); *M*: malonylawobanin (**11**); *F*: flavocommelin, *C*: commelinin; aged; room temperature for a few years. (b) Schematic representation of 7 spots of commelinin and demalonylcommelinins on electrophoresis. [From H. Tamura, T. Kondo, and T. Goto: Tetrahedron Lett. **27** 1801 (1986)]

communis is malonylawobanin (M) (**11**) containing a malonic acid residue (*35*). Awobanin (**12**) is an artifact produced from malonylawobanin
(**11**) during extraction with methanolic hydrochloric acid.

Another mysterious finding is that natural commelinin (aged) gives
seven spots on electrophoresis (*81*) (Fig. 32). Goto *et al.* found that
commelinin prepared from M, F and magnesium ions produced only
the spot which moved fastest, and that commelin synthesized from
A, F and magnesium ion gave only the spot with the least movement
(*38, 104*). When a mixture of M and A was used for synthesis of the
pigment, all seven spots were obtained. Natural commelinin newly pre-

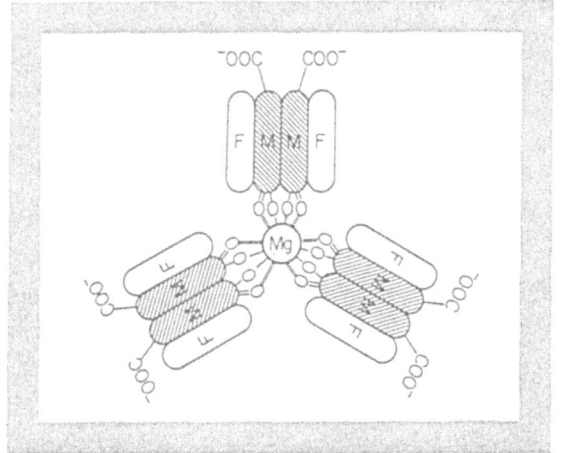

Fig. 33. (a) A proposed structure of a half of commelinin molecule. [From T. Goto
et al.: Ann. New York Acad. Sci. **471**, 155 (1986), with permission]. (b) A proposed
gross structure of commelinin (revised). Magnesium has an octahedral structure. Another
magnesium atom is under the central magnesium

pared from fresh petals of *Commelina communis* was identical with the spot which moved fastest. This result indicated that commelinin must contain six molecules each of malonylawobanin (M) (**11**) and flavocommelin (F) (**34**) per molecule of commelinin, the calculated molecular weight being about 9,500. The value observed by analytical ultracentrifugation was around 10,000. The compositions of these seven spots must be as shown in Fig. 32 (*104*). At least one hydroxyl group is necessary at the 3'-position of the B-ring to form such a complex, indicating that the metal was chelated to this position.

Crystals of commelinin and the cadmium analog of commelinin have been prepared by TAKEDA (*96*) and have been subjected to X-ray crystallographic analysis by Y. SAITO, M. ITO and S. OBA (personal commun.). Although a complete structural analysis of cadmium-commelinin was not successful, the following crystal data were obtained: trigonal, $a = 31.251$ Å, $c = 33.675$ Å, space group P321, $Z = 12(?)$ (AF as one unit), $U = 28.529$ Å^3. This crystal has a three-fold symmetry around the c-axis.

Considering these crystal data and the composition, GOTO *et al.* proposed the gross structure of commelinin as shown in Fig. 33 (*38*). Thus the commelinin molecule has the composition $(M_6F_6Mg_2)^{6-}$, whose structure should have three-fold symmetry around the c-axis. This is fulfilled by putting three M_2F_2 units around the c-axis and two magnesium atoms at the center. Two co-pigmented MF units are stacked chirally by hydrophobic interaction between the two anthocyanin molecules to form the FMMF unit; a large exciton type Cotton

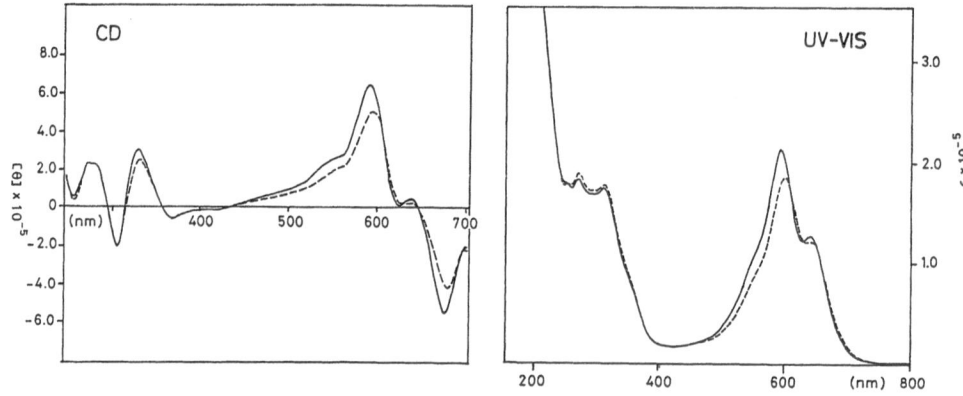

Fig. 34. Electronic and CD spectra of commelinin and demalonylcommelinin in 0.1 M phosphate buffer at pH 7.0 (5×10^{-5} M; 1 mm cell). — Commelinin; --- demalonylcommelinin prepared from awobanin, flavocommelin and magnesium ion. [From H. TAMURA, T. KONDO, and T. GOTO: Tetrahedron Lett. **27**, 1801 (1986)]

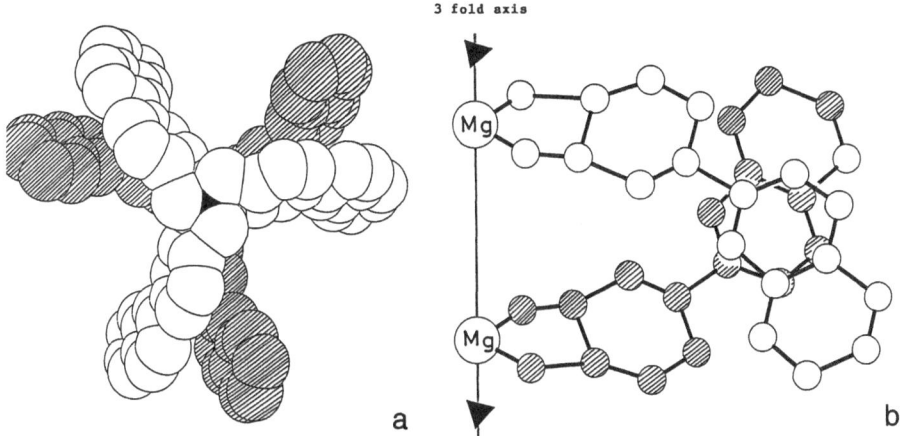

Fig. 35. Proposed stacking pattern of six anthocyanin chromophores consisting of two (anthocyanidin)₃Mg units, which is suggested from a very crude analysis of X-ray crystallographic data of commelinin and Cd-commelinin. (a) Space-filling drawing of anthocyanidin parts through c-axis. Only carbon and oxygen atoms of anthocyanidin parts are presented. The black spot indicates a magnesium atom. Another magnesium atom is under it. The hydrophobic stacking between two aromatic nuclei (self-association of anthocyanin) is apparent. (b) Side view. [T. Goto, T. Kawai, T. Kondo, Y. Saito, M. Ito, S. Oba, K. Takeda, and K. Hayashi, unpublished results]

band indicates the chiral stacking of the two anthocyanins (Fig. 34). Three FMMF units are doubly linked by chelation with two magnesium atoms. A more detailed stacking form of the anthocyanidin nuclei is proposed in Fig. 35. Saito *et al.* assumed a structure for commelinin from the X-ray data (*87*).

7.2 Protocyanin

Willstätter in 1913 (*115*) isolated cyanin (**2**) from cornflower as the anthocyanin component but Tamura *et al.* (*105*) and Takeda and Tominaga (*103*) independently found much later that the true anthocyanin is not cyanin but succinylcyanin (centaurocyanin) (**13**). A flavone (**35**) isolated from cornflower by Asen and Horowitz (*2*) was also found to be an artifact; the true flavone (**36**) contains an additional malonic acid residue (*105*).

From blue cornflower, *Centaurea cyanus*, Bayer (*11*) isolated protocyanin (**37**) as the intact pigment that contained 19.2% cyanin and about 80% of a polysaccharide as well as ferric and aluminum ions (molecular weight about 6,200) (*12*) and proposed a structure, in which

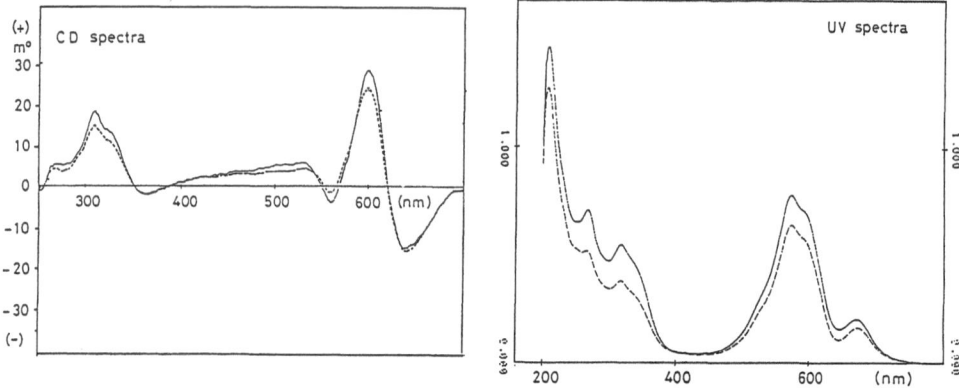

Fig. 36. Electronic and CD spectra of natural and synthetic protocyanin in 0.1 M acetate
buffer at pH 5.0 (520 µg/ml; 1 mm cell). --- Natural; — synthetic

trivalent metal ions coordinatively bind two cyanin molecules and poly-
galacturonic acid. Hayashi *et al.* (*47*) reported that the protocyanin
which they obtained in crystalline form yielded on hydrolysis cyanin,
a flavone, a carbohydrate, and a peptide as well as iron, magnesium,
and potassium ions. Asen and Jurd (*3*) isolated from cornflower a
crystalline blue pigment and concluded that it was an iron complex
of four molecules of cyanin (**2**) and three molecules of a flavone glucoside, the structure of which was later (*2*) shown to be apigenin 4′-0-β-D-
glucoside 7-0-β-D-glucuronide (**35**); no mention was made of the presence of high molecular weight substances. They thought that this pigment differed markedly from protocyanin and named it cyanocentaurin, but later work by Osawa (*80*) led to the conclusion that protocyanin and cyanocentaurin were identical.

Protocyanin like commelinin is not dialyzable and migrates to the
anode on electrophoresis at pH 4–6 (*81*), indicating that it is a high
molecular weight pigment bearing negative charge(s). Goto *et al.* purified the cornflower pigment extensively and obtained protocyanin free
of macromolecular substances such as polysaccharides or polypeptides
(*38*). It was composed of succinylcyanin (**13**), the malonylflavone (**36**),
and Fe and Mg ions. Pigment reconstructed from the above components was identical with the natural pigment which had the following
properties: On electrophoresis, the pigment moved toward the anode
nearly twice as fast as commelinin. It showed a large exciton type
CD curve (Fig. 36). Its molecular weight determined by ultracentrifugation was about 8,500. Thus, the composition of the pigment was sug-
gested to be $[(\text{succinylcyanin}:\text{malonylflavone})_6.\text{Fe.Mg}]^{-12}$ or close to
this, with a gross structure similar to that of commelinin.

References, pp. 153–158

The blue color of *Hydrangea macrophylla* was believed to be due to one of the metalloanthocyanins containing Al (*10*), and recent reports (*100*) suggested that the blue color is due mainly to a delphinidin 3-glucoside-aluminum-3-caffeylquinic acid complex (see Chapter 5).

References

1. ABE, K., Y. SAKAINO, J. KAKINUMA, and H. KAKISAWA: Structures of Anthocyanin Anhydro Bases. Nippon Kagaku Kaishi, 1197–1204 (1977).

2. ASEN, S., and R.M. HOROWITZ: Apigenin 4'-0-β-D-glucoside 7-0-β-D-glucuronide: The Copigment in the Blue Pigment of *Centaurea cyanus*. Phytochem. **13**, 1219–1223 (1974).

3. ASEN, S., and L. JURD: The Constitution of a Crystalline, Blue Cornflower Pigment. Phytochem. **6**, 577–584 (1967).

4. ASEN, S., K.H. NORRIS, and R.N. STEWART: Effect of pH and Concentration of the Anthocyanin-flavonol Co-pigment Complex on the Color of 'Better Times' Roses. J. Am. Soc. Hort. Sci. **96**, 770–773 (1971).

5. ASEN, S., R.N. STEWART, and K.H. NORRIS: Co-pigmentation Effect of Quercetin Glycosides on Absorption Characteristics of Cyanidin Glycosides and Color of Red Wing Azalea. Phytochem. **10**, 171–175 (1971).

6. ASEN, S., R.N. STEWART, and K.H. NORRIS: Co-pigmentation of Anthocyanins in Plant Tissues and its Effect on Color. Phytochem. **11**, 1139–1144 (1972).

7. ASEN, S., R.N. STEWART, and K.H. NORRIS: Anthocyanin, Flavonol Copigments, and pH Responsible for Larkspur Flower Color. Phytochem. **14**, 2677–2682 (1975).

8. ASEN, S., R.N. STEWART, and K.H. NORRIS: Anthocyanin and pH Involved in the Color of 'Heavenly Blue' Morning Glory. Phytochem. **16**, 1118–1119 (1977).

9. ASEN, S., R.N. STEWART, K.H. NORRIS, and D.R. HASSIE: A Stable Blue Non-metallic Co-pigment Complex of Delphanin and C-Glycosylflavones in Prof. Blaauw Iris. Phytochem. **9**, 619–627 (1970).

10. ASEN, S., N.W. STUART, and H.W. SIEGELMAN: Effect of Various Concentrations of Nitrogen, Phosphorus and Potassium on Sepal Color of *Hydrangea macrophylla*. Proc. Am. Soc. Hort. Sci. **73**, 495–502 (1959).

11. BAYER, E.: Über den blauen Farbstoff der Kornblume I. Natürliche und synthetische Anthocyan-Metallkomplexe. Chem. Ber. **91**, 1115–1122 (1958).

12. BAYER, E.: Complex formation and Flower Colors. Angew. Chem. Internat. Edit. **5**, 791–798 (1966).

13. BIRKOFER, L., C. KAISER, and W. KOCH: Konstitution von Acyl-anthocyanen. Z. Naturforsch. **20b**, 424–426 (1965).

14. BIRKOFER, L., C. KAISER, W. KOCH, M. DONIKE, and D. WOLF: Acylierte Anthocyane II. cis-trans-Isomerie bei Acyl-Anthocyanen. Z. Naturforsch. **18b**, 631–634 (1963).

15. BLOOM, M., and T.A. GEISSMAN: Malonic acid: the Acyl Moiety of the *Minulus leteus* Anthocyanin. Phytochem. **12**, 2005–2006 (1973).

16. BRIDLE, P., R.S.T. LOEFFLER, C.F. TIMBERLAKE, and R. SELF: Cyanidin 3-Malonylglucoside in *Cichorium intybus*. Phytochem. **23**, 2968–2969 (1984).

17. BROUILLARD, R.: Origin of the Exceptional Colour Stability of the *Zebrina* Anthocyanin. Phytochem. **20**, 143–145 (1981).

18. BROUILLARD, R.: Chemical Structure of Anthocyanins. In: Anthocyanins as Food Colors, Chapter 1, pp 1–40, Academic Press, New York, 1982.

19. BROUILLARD, R.: The *in vivo* Expression of Anthocyanin Colour in Plants. Phytochem. **22**, 1311–1323 (1983).

20. BROUILLARD, R., and B. DELAPORTE: Chemistry of Anthocyanin Pigments. 2. Kinetic and Thermodynamic Study of Proton Transfer, Hydration, and Tautomeric Reactions of Malvidin 3-Glucoside. J. Am. Chem. Soc. **99**, 8461–8468 (1977).
21. BROUILLARD, R., and J.-E. DUBOIS: Mechanism of the Structural Transformations of Anthocyanins in Acidic Media. J. Am. Chem. Soc. **99**, 1359–1364 (1977).
22. BROUILLARD, R., G.A. IACOBUCCI, and J.G. SWEENY: Chemistry of Anthocyanin Pigments. 9. Uv-Visible Spectrophotometric Determination of the Acidity Constants of Apigeninidin and Three Related 3-Deoxyflavylium Salts. J. Am. Chem. Soc. **104**, 7585–7590 (1982).
23. CHEMINAT, A., and R. BROUILLARD: PMR Investigation of 3-0-(β-D-glucosyl)malvidin. Structural Transformation in Aqueous Solutions. Tetrahedron Lett. **27**, 4457–4460 (1986).
24. CHEN, L.-J., and G. HRAZDINA: Structural Aspects of Anthocyanin-flavonoid Complex Formation and its Role in Plant Color. Phytochem. **20**, 297–303 (1981).
25. CORNUZ, G., H. WYLER, and J. LAUTERWEIN: Pelargonidin 3-malonylsophoroside from the Red Iceland Poppy, *Papaver nudicaule*. Phytochem. **20**, 1461–1462 (1981).
26. EVEREST, A.E., and A.J. HALL: Anthocyanins and Anthocyanidins IV. Observations on: (a) Anthocyan Colors in Flowers, and (b) The Formation of Anthocyans in Plants. Proc. Roy. Soc. London **92 B**, 150–162 (1921).
27. GOTO, T., T. HOSHINO, and M. OHBA: Stabilization Effect of Neutral Salts on Anthocyanins, Flavylium Salts, Anhydro Bases and Genuine Anthocyanins. Agric. Biol. Chem. **40**, 1593–1596 (1976).
28. GOTO, T., T. HOSHINO, and S. TAKASE: 20th Symposium on the Chemistry of Natural Products. Symposium Papers pp 59–66 (1976).
29. GOTO, T., T. HOSHINO, and S. TAKASE: A Proposed Structure of Commelinin, a Sky-blue Anthocyanin Complex Obtained from the Flower Petals of *Commelina*. Tetrahedron Lett. 2905–2908 (1979).
30. GOTO, T., H. IMAGAWA, and T. KONDO: Heavenly Blue Anthocyanin III. Structure of Bis-deacyl Heavenly Blue Anthocyanin, a Controlled Alkaline Hydrolysis Product of Heavenly Blue Anthocyanin. Heterocycles **17**, 355–358 (1982).
31. GOTO, T., T. KONDO, H. IMAGAWA, and I. MIURA: Heavenly Blue Anthocyanin II. *Trans*-4-O-(6-*trans*-3-O-(β-D-glucopyranosyl)caffeyl)-β-D-glucopyranosyl)caffeic Acid, a Novel Component in Heavenly Blue Anthocyanin. Tetrahedron Lett. **22**, 3213–3216 (1981).
32. GOTO, T., T. KONDO, H. IMAGAWA, S. TAKASE, M. ATOBE, and I. MIURA: Structure Confirmation of Tris-Deacyl Heavenly Blue Anthocyanin obtained from Flower of Morning Glory "Heavenly Blue". Chemistry Lett. 883–886 (1981).
33. GOTO, T., T. KONDO, H. TAMURA, H. IMAGAWA, A. IINO, and K. TAKEDA: Structure of Gentiodelphin, an Acylated Anthocyanin Isolated from *Gentiana makinoi*, That Is Stable in Dilute Aqueous Solution. Tetrahedron Lett. **23**, 3695–3698 (1982).
34. GOTO, T., T. KONDO, H. TAMURA, and K. KAWAHORI: Structure of Platyconin, a Diacylated Anthocyanin Isolated from the Chinese Bellflower *Platycodon grandiflorum*. Tetrahedron Lett. **24**, 2181–2184 (1983).
35. GOTO, T., T. KONDO, H. TAMURA, and S. TAKASE: Structure of Malonylawobanin, the Real Anthocyanin Present in Blue-Colored Flower Petals of *Commelina communis*. Tetrahedron Lett. **24**, 4863–4866 (1983).
36. GOTO, T., T. KONDO, H. KAWAI, and H. TAMURA: Structure of Cinerarin, a Tetraacylated Anthocyanin Isolated from the Blue Garden Cineraria, *Senecio cruentus*. Tetrahedron Lett. **25**, 6021–6024 (1984).
37. GOTO, T., S. TAKASE, and T. KONDO: PMR Spectra of Natural Acylated Anthocyanins. Determination of Stereostructure of Awobanin, Shisonin and Violanin. Tetrahedron Lett. 2413–2416 (1978).

38. GOTO, I., H. TAMURA, T. KAWAI, T. HOSHINO, N. HARADA, and T. KONDO: Chemistry of Metalloanthocyanins. Annals New York Acad. Sci. **471**, 155–173 (1986).
39. GOTO, T., H. TAMURA, and T. KONDO: Unpublished.
40. HARBORNE, J.B.: Spectral Methods of Characterizing Anthocyanins. Biochem. J. **70**, 22–28 (1958).
41. HARBORNE, J.B.: Plant Polyphenols-XI. The Structure of Acylated Anthocyanins. Phytochem. **3**, 151–160 (1964).
42. HARBORNE, J.B.: Flavonoids: Distribution and Contribution to Plant Colour. In: T.W. GOODWIN (ed.) Chemistry and Biochemistry of Plant Pigments, Chapter 9, pp 247–267. Academic Press, New York, 1965.
43. HARBORNE, J.B.: Comparative Biochemistry of the Flavonoids. Academic Press, London, 1967.
44. HARBORNE, J.B., and M. BOARDLEY: The Widespread Occurrence in Nature of Anthocyanins as Zwitterions. Z. Naturforsch. **40c**, 305–308 (1985).
45. HARPER, K.A.: Structure Changes of Flavylium Salts. J. Biol. Chem. Biothechnol. **23**, 261–271 (1973).
46. HAYASHI, K., Y. ABE, and S. MITSUI: Blue Anthocyanin from the Flowers of *Commelina*, the Crystallization and Some Properties Thereof. Studies on Anthocyanins XXX. Proc. Japan Acad. **34**, 373–378 (1958).
47. HAYASHI, K., N. SAITO, and K. MITSUI: Anthocyanins (XXIV) Metallic Components in Newly Crystallized Specimens of Bayer's Protocyanin, a Blue Metallo-Anthocyanin from the Cornflower. Proc. Japan Acad. **37**, 393–397 (1961).
48. HAYASHI, K., and K. TAKEDA: Further Purification and Component Analysis of Commelinin Showing the Presence of Magnesium in This Blue Complex Molecule. Studies on Anthocyanins LXII. Proc. Japan Acad. **46**, 535–540 (1970).
49. HOSHINO, T., U. MATSUMOTO, and T. GOTO: Evidences of the Self-association of Anthocyanins I. Circular Dichroism of Cyanin Anhydrobase. Tetrahedron Lett. **21**, 1751–1754 (1980).
50. HOSHINO, T., U. MATSUMOTO, and T. GOTO: The Stabilizing Effect of the Acyl Group on the Co-pigmentation of Acylated Anthocyanins with C-Glucosylflavones. Phytochem. **19**, 663–667 (1980).
51. HOSHINO, T., U. MATSUMOTO, and T. Goto: Self-association of Some Anthocyanins in Neutral Aqueous Solution. Phytochem. **20**, 1971–1976 (1981).
52. HOSHINO, T., U. MATSUMOTO, T. GOTO, and N. HARADA: Evidence for the Self-association of Anthocyanins IV. PMR Spectroscopic Evidence for the Vertical Stacking of Anthocyanin Molecules. Tetrahedron Lett. **23**, 433–436 (1982).
53. HOSHINO, T., U. MATSUMOTO, N. HARADA, and T. GOTO: Chiral Exciton Coupled Stacking of Anthochanins: Interpretation of the Origin of Anomalous CD Induced by Anthocyanin Association. Tetrahedron Lett. **22**, 3621–3624 (1981).
54. HRAZDINA, G.: Anthocyanins. In: J.B. HARBORNE and T. MABRY (ed.), The Flavonoids, Chapter 3, pp 135–186, Chapman & Hall, London, 1982.
55. HRAZDINA, G., H. IREDALE, and L.R. MATTICK: Anthocyanin Composition of *Brassica oleracea* cv. Red Danish. Phytochem. **16**, 297–299 (1977).
56. IACOBUCCI, G., and J.G. SWEENY: The Chemistry of Anthocyanins, Anthocyanidins and Related Flavylium Salts. (review) Tetrahedron **39**, 3005–3038 (1983).
57. IDAKA, E., Y. OHASHI, T. OGAWA, T. KONDO, and T. GOTO: Structure of Zebrinin, a Novel Acylated Anthocyanin Isolated From *Zebrina pendula*. Tetrahedron Lett., **28**, 1901–1904 (1987).
58. IMBERT, M.P., C.E. SEAFORTH, and D.B. WILLIAMS: Anthocyanin Pigments of the Sweet Potato, *Ipomoea batatas*. Proc. Am. Soc. Hort. Sci. **88**, 481–485 (1966).
59. ISHIKURA, N., and S. HAYASHIDA: Kaempferol Glycosides in the Seedcoat of *Ophiopogon jaburan* (Kunth) Lodd. Agric. Biol. Chem. **43**, 1923–1926 (1979).
60. ISHIKURA, N., and M. SHIMIZU: Studies on the Flower Color of Morning Glory I.

Anthocyanins and Some Concomitant Phenolics in the Blue Petals of *Ipomoea ruburo-coerulea* Hook, "Heavenly Blue". Kumamoto J. Sci. Biol. **12**, 41–70 (1975).

61. ISHIKURA, N., and U. TAKAHAMA: Notes on the Anthocyanidin Types and Metals in Blue Flowers and Fruits. Kumamoto J. Sci. Biol. **11**, 13–17 (1972).

62. ISHIKURA, N., and E. YAMAMOTO: Some Factors Involved in the Blue Color of 'Heavenly Blue' Morning Glory Flowers (in Japanese). Nippon Nogeikagaku Kaishi **54**, 637–643 (1980).

63. ISHIKURA, N., and K. YOSHITAMA: Anthocyanin-Flavonol Co-pigmentation in Blue Seed Coats of *Ophiopogon jaburan*. J. Plant Physiol. **115**, 171–175 (1984).

64. JURD, L.: Anthocyanins and Related Compounds. I. Structural Transformations of Flavylium Salts in Acidic Solutions. J. Org. Chem. **28**, 987–991 (1963).

65. JURD, L., and T.A. GEISSMAN: Anthocyanins and Related Compounds. II. Structure Transformations of Some Anhydro Bases. J. Org. Chem. **28**, 2394–2397 (1968).

66. KARRER, P., and R. WIDMER: Pflanzenfarbstoffe VIII. Ueber die Konstitution des Monardaeins. Helv. Chim. Acta **11**, 837–842 (1928).

67. KARRER, P., and R. WIDMER: Zur Konstitution des Monardaeins und Salvianins. XII. Mitteilung über Pflanzenfarbstoffe. Helv. Chim. Acta **12**, 292–295 (1929).

68. KONDO, T., T. KAWAI, H. TAMURA, and T. GOTO: Structure Determination of Heavenly Blue Anthocyanin, a Complex Monomeric Anthocyanin from the Morning Glory *Ipomoea tricolor*, by Means of the Negative NOE Method. Tetrahedron Lett., **28**, 2273–2276 (1987).

69. KONDO, T., Y. NAKANE, H. TAMURA, T. GOTO, and G.H. EUGSTER: Structure of Monardaein, a bis-Malonylated Anthocyanin Isolated from Golden Balm, *Monarda didyma*. Tetrahedron Lett. **26**, 5879–5882 (1985).

70. KONDO, T., H. TAMURA, S. TAKASE, and T. GOTO: Complete Stereostructure Determination of Natural Acylated Anthocyanins, Violanin, Awobanin and Malonylawobanin by ¹H-NMR. Nippon Kagaku Kaishi 1571–1578 (1986).

71. KURODA, C.: The Colouring Matter of "Awobana". Part I. Proc. Imp. Acad. **7**, 61–63 (1931).

72. KURODA, C.: The Colouring Matter of "Awobana". Part II. Proc. Imp. Acad. **9**, 94–96 (1933).

73. KURODA, C.: The Colouring Matter of "Awobana" Part III. Proc. Imp. Acad. **11**, 238–239 (1935).

74. KURODA, C.: The Constitution of Awobanin and Awobanol, the Colouring Matter of Awobana and its Copigment. Bull. Chem. Soc. Japan **11**, 265–270 (1936).

75. MCCLELLAND, R.A., and S. GEDGE: Hydration of the Flavylium Ion. J. Am. Chem. Soc. **102**, 5838–5848 (1980).

76. MITSUI, S., K. HAYASHI, and S. HATTORI: Further Studies on Commelinin, a Crystalline Blue Metalloanthocyanin from the Flowers of *Commelina*. Studies on Anthocyanins XXXI. Proc. Japan Acad. **35**, 169–174 (1959).

77. MITSUI, S., K. HAYASHI, and S. HATTORI: Crystallization and Properties of Commelinin, a Blue Metallo-anthocyanin from *Commelina*. (Studies on Anthocyanins XXXI) (in Japanese). Bot. Mag. Tokyo **72**, 853–854 (1959).

78. NILSSON, E.: Studies of Flavylium Compounds. Chemica Scripta **4**, 49–55 (1973).

79. OOTANI, S.: Chemico-genetical Studies on the Flower Colors and Pigment Components in Wild and Cultivated Pansies, with Special Regard to their Inter-varietal Relationship. Rep. Inst. for Breeding Res., Tokyo Univ. Agric., Supplement 1–72 (1973).

80. OSAWA, Y.: Copigmentation of Anthocyanins. In: P. MARKAKIS (ed.), Anthocyanins as Food Colors, Chapter 2, pp 41–68. Academic Press, New York, 1982.

81. OSAWA, Y., M. KOIZUMI, and N. SAITO: Zone Electrophoretic Investigation of Anthocyanins. Phytochem. **10**, 1591–1593 (1971).

82. POMILIO, A.B., and J.F. SPROVIERO: Complex Anthocyanins From *Ipomoea congesta*. Phytochem. **11**, 2323–2326 (1972).

83. ROBINSON, G.M., and R. ROBINSON: CLXXXII. A Survey of Anthocyanins. I. Biochem. J. 25, 1687–1705 (1931).
84. ROBINSON, R., and G.M. ROBINSON: The Colloid Chemistry of Leaf and Flower Pigments and the Precursors of the Anthocyanins. J. Am. Chem. Soc. 61, 1605–1606 (1939).
85. SAITO, N., K. ABE, T. HONDA, C.F. TIMBERLAKE, and P. BRIDLE: Acylated Delphinidin Glucosides and Flavonols from Clitoria ternatea. Phytochem. 24, 1583–1586 (1985).
86. SAITO, N., Y. OSAWA, and K. HAYASHI: Platyconin, a New Acylated Anthocyanin in Chinese Bell-Flower, Platycodon grandiflorum. Phytochem. 10, 445–447 (1971).
87. SAITO, N., K. UENO, and K. TAKEDA: The Crystallographic Study of Commelinin, a Blue Metalloanthocyanin, and Its Reconstituted Analogues by the X-ray Diffraction Method. Meijigakuin Daigaku Ippankyoiku Fuzokukenkyusho Kiyo No 10, 63–81 (1986).
88. SAITO, N., Y. OSAWA, and K. HAYASHI: Isolation of a Blue-Violet Pigment from the Flowers of Platycodon grandiflorum. Bot. Mag. Tokyo 85, 105–110 (1972).
89. SAITO, N., C.F. TIMBERLAKE, O.G. TUCKNOTT, and I.A.S. LEWIS: Fast Atom Bombardment Mass Spectrometry of the Anthocyanins Violanin and Platyconin. Phytochem. 22, 1007–1009 (1983).
90. SCHEFFELDT, P., and G. HRAZDINA: Co-pigmentation of Anthocyanins under Physiological Conditions. J. Food Science 43, 517–520 (1978).
91. SHIBATA, K., Y. SHIBATA, and I. KASIWAGI: Studies on Anthocyanin. Color Variation in Anthocyanins. J. Am. Chem. Soc. 41, 208–220 (1919).
92. SINGLETON, V.L.: Common Plant Phenols Other than Anthocyanins, Contributions to Coloration and Discoloration. In: C.O. CHICHESTER (ed.) The Chemistry of Plant Pigments, pp 143–191. Academic Press, New York, 1972.
93. SOMMERS, T.C., and M.E. EVANS: Grape Pigment Phenomena: Interpretation of Major Colour Losses During Vinification. J. Sci. Food Agric. 30, 623–633 (1979).
94. SULYOK, G., and A. LASZLO-BENCSIK: Cyanidin 3-(6-succinylglucoside)-5-glucoside from Flowers of Seven Centaurea Species. Phytochem. 24, 1121–1122 (1985).
95. SWEENY, J.G., M.M. WILKINSON, and G.A. IACOBUCCI: Effect of Flavonoid Sulfonates on the Photobleaching of Anthocyanins in Acid Solution. J. Agric. Food. Chem. 29, 563–567 (1981).
96. TAKEDA, K.: Metallo-anthocyanins II. Further Experiments of Synthesizing Crystalline Blue Metallo-anthocyanins Using Various Kinds of Bivalent Metals. Proc. Japan Acad. 53, Ser. B, 257–261 (1977).
97. TAKEDA, K., T. FUJII, and M. IIDA: Magnesium in the Blue Pigment Complex Commelinin. Phytochem. 23, 879–881 (1984).
98. TAKEDA, K., and K. HAYASHI: Oxidative Degradation of Acylated Anthocyanins Showing the Presence of Organic Acid-sugar Linkage in the 3-Position of Anthocyanidins; Experiments on Ensatin, Awobanin, and Shisonin. Studies on Anthocyanins XLIII. Proc. Japan Acad. 40, 510–515 (1964).
99. TAKEDA, K., and K. HAYASHI: Metallo-anthocyanins I. Reconstruction of Commelinin from Its Components, Awobanin, Flavocommelin and Magnesium. Proc. Japan Acad. 53, Ser B, 1–5 (1977).
100. TAKEDA, K., R. KUBOTA, and C. YAGIOKA: Copigments in the Blueing of Sepal Colour of Hydrangea macrophylla. Phytochem. 24, 1207–1209 (1985); K. TAKEDA, M. KARIUDA, and H. ITOI: Blueing of Sepal Colour of Hydrangea macrophylla. Phytochem. 24, 2251–2254 (1985).
101. TAKEDA, K., S. MITSUI, and K. HAYASHI: Structure of a New Flavonoid in the Blue Complex Molecule of Commelinin. Studies on Anthocyanins LIV. Bot. Mag. Tokyo 79, 578–587 (1966).
102. TAKEDA, K., N. SAITO, and K. HAYASHI: Further Experiments on the Structure

of Genuine Anthocyanins. Studies on Anthocyanins LIX. Proc. Japan Acad. **44**, 352–357 (1968).

103. TAKEDA, K., and S. TOMINAGA: The Anthocyanin in Blue Flowers of *Centaurea cyanus*. Bot. Mag. Tokyo **96**, 359–363 (1983).
104. TAMURA, H., T. KONDO, and T. GOTO: The Composition of Commelinin, a Highly Associated Metalloanthocyanin Present in the Blue Flower Petals of *Commelina communis*. Tetrahedron Lett. **27**, 1801–1804 (1986).
105. TAMURA, H., T. KONDO, Y. KATO, and T. GOTO: Structure of Succinyl Anthocyanin and a Malonyl Flavone, Two Constituents of the Complex Blue Pigment of Cornflower *Centaurea cyanus*. Tetrahedron Lett. **24**, 5749–5752 (1983).
106. TANCHEV, S.S., and C.F. TIMBERLAKE: The Anthocyanins of Red Cabbage (*Brassica oleracea*). Phytochem. **8**, 1825–1827 (1969).
107. TIMBERLAKE, C.F.: Anthocyanins – Occurrence, Extraction and Chemistry. Food Chem. **5**, 69–80 (1980).
108. TIMBERLAKE, C.F.: Anthocyanins in Fruits and Vegetables. In: J. FRIEND, and M.J.C. RHODES (ed.), Recent Advances in the Biochemistry of Fruit and Vegetables, Chapter 12, pp 221–247. Academic Press, New York, 1981–1982.
109. TIMBERLAKE, C.F., and P. BRIDLE: Spectral Studies of Anthocyanin and Anthocyanidin Equilibria in Aqueous Solution. Nature **212**, 158–159 (1966).
110. TIMBERLAKE, C.F., and P. BRIDLE: The Anthocyanins. In: J.B. HARBORNE, T.J. MABRY, and H. MABRY (ed.), The Flavonoids, Chapter 5, pp 215–266. Chapman & Hall, London, 1975.
111. TIMBERLAKE, C.F., and P. BRIDLE: Anthocyanins. In: J. WALFORD (ed.), Developments in Food Colours – 1, Chapter 5, pp 115–150. Applied Science Publ., London, 1980.
112. WATANABE, S., S. SAKAMURA, and Y. OBATA: The Structure of Acylated Anthocyanins in Eggplant and Perilla, and the Position of Acylation. Agric. Biol. Chem. **30**, 420–422 (1966).
113. WILLIAMS, M., and G. HRAZDINA: Anthocyanins as Food Colorants: Effect of pH on the Formation of Anthocyanin-Rutin Complexes. J. Food Science **44**, 66–68 (1979).
114. WILLSTÄTTER, R., and E.K. BOLTON: Untersuchungen über die Anthocyane; XI. Über das Anthocyan der rotblühenden Salviaarten. Liebigs Ann. Chem. **412**, 113–137 (1917).
115. WILLSTÄTTER, R., and A.E. EVEREST: Untersuchungen über die Anthocyane I. Über den Farbstoff der Kornblume. Liebigs Ann. Chem. **401**, 189–232 (1913).
116. YAZAKI, Y.: Co-pigmentation and the Color Change with Age in Petals of *Fuchsia hybrida*. Bot. Mag. Tokyo **89**, 45–57 (1976).
117. YAZAKI, Y., and K. HAYASHI: Analysis of Flower Colors in *Fuchsia hybrida* in Reference to the Concept of Co-pigmentation. Studies on Anthocyanins. LVI. Proc. Japan Acad. **43**, 316–321 (1967).
118. YOSHITAMA, K.: An Acylated Delphinidin 3-Rutinoside-5,3′,5′-triglucoside from *Lobelia erinus*. Phytochem. **16**, 1857–1858 (1977).
119. YOSHITAMA, K.: Blue and Purple Anthocyanins Isolated from the Flowers of *Tradescantia reflexa*. Bot. Mag. Tokyo **91**, 207–212 (1978).
120. YOSHITAMA, K., and K. HAYASHI: Concerning the Structure of Cinerarin, a Blue Anthocyanin From Garden Cineraria. Studies on Anthocyanins, LXVI. Bot. Mag. Tokyo **87**, 33–40 (1974).
121. YOSHITAMA, K., K. HAYASHI, K. ABE, and H. KAKISAWA: Further Evidence for the Glycoside Structure of Cinerarin. Studies on Anthocyanins, LXVII. Bot. Mag. Tokyo **88**, 213–217 (1975).

(*Received February 19, 1987*)

Carbazole Alkaloids

By P. BHATTACHARYYA and D.P. CHAKRABORTY
Bose Institute, Calcutta, India

Contents

I. Introduction

Naturally occurring carbazole alkaloids were reviewed first by Kapil (*38*) in 1971 and then by one of us (*23*) in 1977. Since then there
has been considerable progress in this field and our present purpose
is to describe further advances in the field and to update the earlier
review.

A. Occurrence

Carbazole alkaloids were reported earlier from higher plants belonging to the genera *Murraya*, *Glycosmis* and *Clausena* of the family
Rutaceae. A recent addition to these genera is *Micromelum* (*21*) from
which a carbazole alkaloid has been reported. A carbazole alkaloid
has also been isolated from the family *Meliaceae* (*40*) which unlike
the Rutaceae is a poor source of nitrogen containing metabolites. Carbazole alkaloids have also been found in a blue alga (*22*) and in a

Streptomyces sp. (*66*). Carbazoles have also been found to be present in cigarette smoke (*72*).

B. Detection of Carbazoles by Chromatographic Methods

In addition to several spray reagents described earlier (*23*) for the detection of carbazoles on a developed chromatogram, concentrated HCl (*64*) has also been used as a spray reagent. Benzoyl peroxide (*16*) has recently been found to be a more effective spray reagent than other previously used reagents. C_{13}-, C_{18}-, and C_{23}-carbazole alkaloids have been separated by gas liquid chromatography (*27*) using 3% OV-17 and 3% SE-30 columns. The OV-17 column requires a higher elution temperature than SE-30 column due to its slightly more polar nature. CHOWDHURY *et al.* (*28*) utilised HPLC technique for the separation of some carbazole alkaloids using chloroform: hexane (30:70) as mobile phase.

II. Methods of Structure Elucidation

A. Physical Methods

1. ^{13}C NMR Spectra

^{13}C nuclear magnetic resonance spectra of carbazole alkaloids have been found to be very helpful in structure elucidation. Very recently several analyses of the ^{13}C spectra of carbazoles and carbazole alkaloids has been published (*1, 9, 44*). Signal identification in the ^{13}C nmr spectra is accomplished mainly by chemical shift comparison with model compounds. The aromatic carbons of carbazole show signals in the region δ 110–140, the aromatic methyl carbon peaks in the region δ 12.6–21.0 and the methoxy carbons in the region δ 55.5–61.4. Chemical shifts in the ^{13}C nmr spectra of some carbazole alkaloids are shown in Table 1.

B. Synthesis of Carbazoles

Only newer methods for the synthesis of carbazoles are briefly summarised below.

P. BHATTACHARYYA and D.P. CHAKRABORTY:

Table 1. ^{13}C NMR Data of Some Carbazole Alkaloids

Carbon No.	Mukonal (50)	2-hy-droxy-3-methyl (7)	Gly-cozo-line (86)	Carb-azo-mycin A (53)	Carb-azo-mycin B (54)	Mup-amine (107)	N-Me Ekeber-ginine (120)	Koeno-line (63)
1	96.3	96.0	110.6	114.4	110.0	104.7	145.5	146.5
2	160.0	154.7	119.8	128.7	127.0	149.8	105.5	105.5
3	115.7	116.2	126.5	144.4	142.0	118.7	125.7	132.6
4	124.9	121.2	126.6	145.9	138.5	121.3	136.3	111.5[a]
5	125.5	120.7	114.5	122.5	122.7	112.1	122.9	120.3
6	119.8[a]	118.2	152.8	119.4	119.5	119.8	120.2	119.2
7	119.9[a]	123.0	102.9	125.0	124.7	104.9	125.6	125.5
8	111.1	110.3	111.4	110.3	110.0	145.6	109.2	110.9[a]
10	145.8	139.5	138.7	136.4	136.8	134.3	134.3	129.2
11	116.9	114.9	122.6	113.5	109.3	117.2	122.7	123.8[b]
12	122.9	120.9	122.6	122.8	123.3	125.0	123.0	123.3[b]
13	140.7	139.7	134.9	139.4	139.3	117.4	141.8	139.3
1'						129.3	26.5	
2'						117.3	122.2	
3'						75.9	132.7	
4'						27.6	25.6	
5'						27.6	18.3	
CHO	193.0						190.1	
OMe			55.5	61.1, 60.5	61.4	55.5	55.6	55.3
CH$_3$		16.64	21.0	12.6, 13.6	12.7, 13.1			
NMe							32.2	
CH$_2$OH								66.2

[a, b] Values may interchange.

1. Palladium Promoted Cyclisation of Diphenylamine

Diphenylamine (1) when heated in acetic acid solution containing palladium acetate furnished carbazole (2) in high yield (2).

(1)   (2)

The method has been utilised for the synthesis of various substituted carbazoles from substituted diphenylamines.

2. Acid Catalysed Cyclisation of β-Ketosulphoxides

Various tetrahydrocarbazoles have been synthesised by this method
(54). β-Ketosulphoxide (3) on treatment with trichloroacetic acid in
boiling dichloroethane readily forms 1-methylthio-2-oxo-1, 2,3,4-tetra-
hydrocarbazole (4). The cyclisation was found to proceed smoothly
upon treatment with toluenesulphonic acid in THF.

(3)   (4)

(5)

Compound (3) on treatment with toluenesulphonic acid in boiling acetonitrile furnished 2-hydroxycarbazole (5). 1-(3-Indolylmethyl)ethyl-
methylsulphinyl methyl ketone (6) in acetonitrile furnished 2-hydroxy-
3-methylcarbazole (7).

(6)   (7)

3. In situ Vinyl Indole Synthesis of Carbazoles

Carbazoles have been synthesised (53) in three steps from indole
(8). Condensation of indole (8) with excess ketone (9) as solvent and
maleic acid (10) as catalyst on reflux furnished tetrahydrocarbazole-2-
carboxylic acid (11). The methyl (or ethyl) ester (12) of the acid was
then dehydrogenated either with chloranil in refluxing o-xylene or

3–10% palladium-on carbon in refluxing o-dichlorobenzene to the corresponding carbazole-2-carboxylate ester (13).

(8)   (9)   (10)   (11)

(13)   (12)

4. Preparation by Application of the Nenitzescu Synthesis

Condensation of p-benzoquinone with anilines containing various electron withdrawing groups furnished carbazole (7). By application of this method 6-hydroxycarbazole derivative (14) has been synthesised from p-benzoquinone (15) and p-nitroaniline (16). Two mechanisms have been proposed for this reaction involving either initial nitrogen-carbon condensation by attack of the amino group on the carbonyl of the benzoquinone or initial attack of the carbon *ortho* to the amino group on the 2-carbon of the benzoquinone:

(15)   (16)   (14)

5. Thermal Cyclisation

Heating 2-(1-cyclohexenyl)-3-(β-methoxyvinyl)-indole (17) in decalin in the presence of 5% Pd-C furnished 1,2,3,4-tetrahydro carbazoles (18, 19) (37).

(17)   **(18)**   **(19)**

6. Free Radical Cyclisation of Diphenylamine to Carbazole

Diphenylamine (1) was cyclised to carbazole in refluxing chloroform with benzoyl peroxide in the presence of light (17). By this method 3-methylcarbazole (21) was also synthesised from the diphenylamine derivative (20).

(1) R=H   (2) R=H
(20) R=CH₃   (21) R=CH₃

Chloroform was found to be the most suitable solvent for the reaction. It is assumed that a CCl₃ radical is first generated under the reaction conditions and that the reaction probably proceeds *via* the intermediate π-radical (22).

$$C_6H_5COO^\bullet + CHCl_3 \longrightarrow C_6H_5COOH + CCl_3^\bullet$$

(22)   **(2)**

7. Intramolecular Diels-Alder Cycloaddition of Vinylketenimines

By intramolecular Diels-Alder cycloaddition of acetylenic vinylke-tenimines, various carbazole derivatives have been synthesized in good yield (30). Thus the carbazole derivative (23) was synthesised from the vinylketenimine (24) generated *in situ* by reaction of anilide (25) with Ph₃PBr₂ in presence of triethylamine in refluxing dichloromethane.

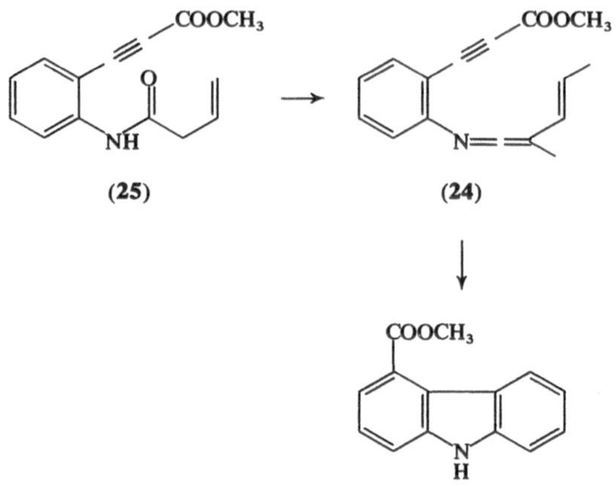

(25)   (24)

(23)

This method has been utilised in the synthesis of N-methyl-tetra-hydroellipticine.

8. Diels-Alder Reaction of Indole-2,3-quinodimethane

A convenient method for the synthesis of carbazole involving the Diels-Alder reaction of an indole-2,3-quino-dimethane intermediate has been developed by SAROJA et al. (67). The N-sodioderivative of 2,3-di-methylindole (26) on benzoylation formed (27) which on bromination furnished N-benzoyl-2,3-dibromomethylindole (28). This on treatment with sodium iodide in DMF in the presence of a suitable dienophile, e.g. dimethylacetylene dicarboxylate, furnished a carbazole derivative (29).

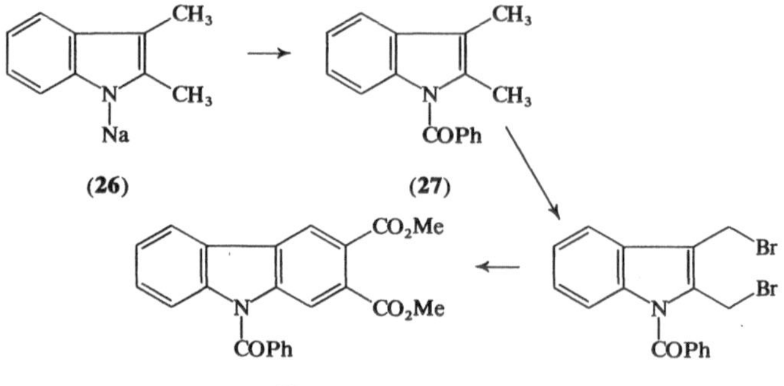

(26)   (27)

(29)   (28)

9. Synthesis of Carbazoles via 2-Vinylindoles

2-(2-Methylpropenyl)-indole (30) on heating with Vilsmeier reagent (DMF/POCl$_3$) furnished 2-methylcarbazole (31) in good yield (6). Similarly 1,2-dimethylcarbazole (33) was obtained from 2-(1,2-dimethylpropenyl)-indole (32).

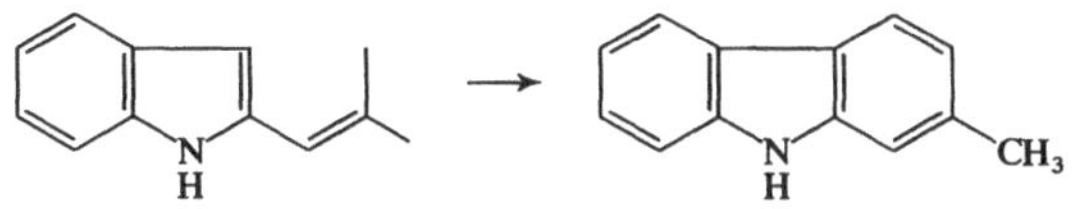

(30)   (31)

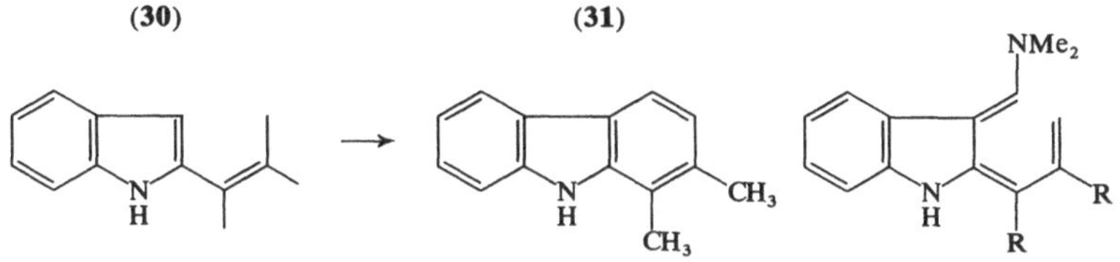

(32)   (33)   A

The reaction probably takes place by electrocyclic ring closure of a hexatriene intermediate of type A.

10. Synthesis of Pyranocarbazole Alkaloids of the C$_{18}$-Carbon Skeleton Group

In recent years various new methods have been developed for the synthesis of pyranocarbazole alkaloids of the C$_{18}$-carbon skeleton group.

(a) Girinimbine (34) has been synthesised in one step by BHATTA-CHARYYA et al. (10) by reaction of 2-hydroxy-3-methylcarbazole (7) with 3-chloro-3-methyl-1-butyne (35) in the presence of AlCl$_3$.

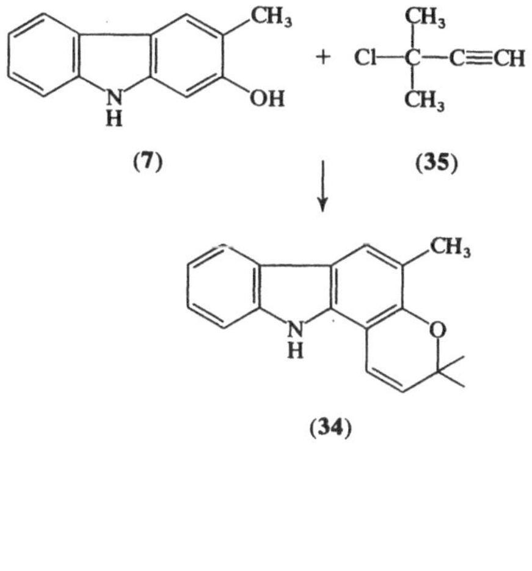

(7)   (35)

(34)

(b) 6-Methoxyheptaphylline (36) has been synthesised from lansine (37) by condensation of 2-methyl-3-buten-2-ol (38) in the presence of boron trifluoride etherate (68)

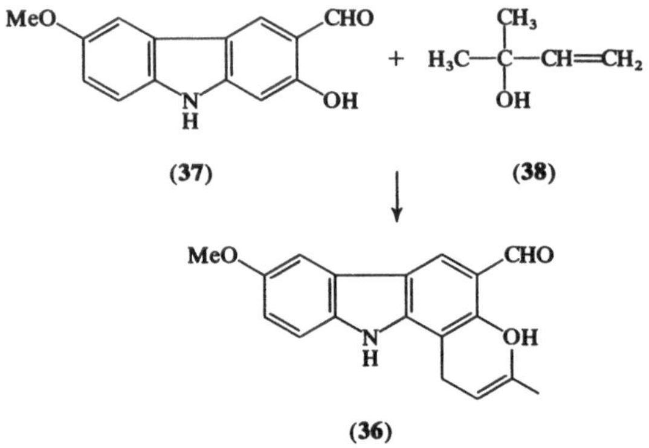

(37)   (38)

(36)

(c) Sʜᴀʀᴍᴀ and Kᴀᴘɪʟ (69) have synthesised heptazoline (39) from the biphenyl derivative (40). Cyclisation of (40) with triethyl phosphite gave 2,8-dimethoxy-3-methylcarbazole (41) whose DDQ oxidation

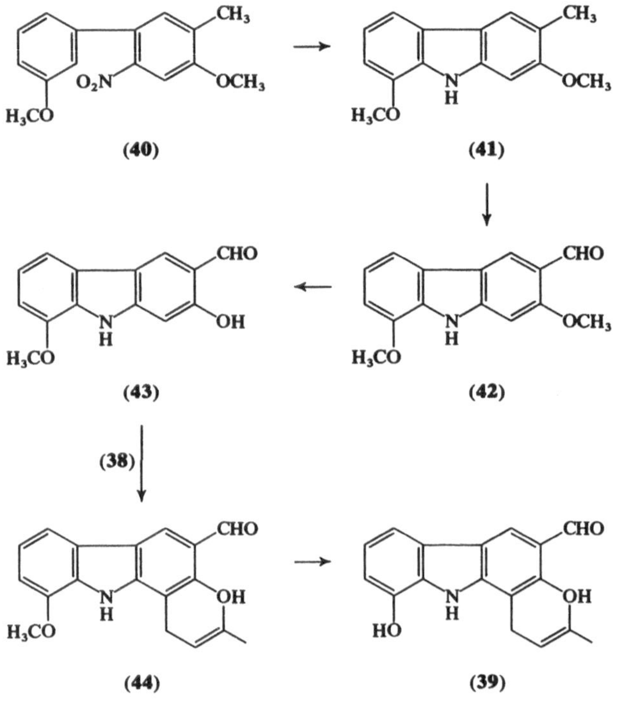

(40)   (41)

(43)   (42)

(44)   (39)

furnished 2,8-dimethoxy-3-formylcarbazole (**42**). Selective demethylation of (**42**) with BCl_3 gave (**43**) which on condensation with 2-methyl-3-butene-2-ol gave (1-(3,3-dimethyl allyl)-2-hydroxy-3-formyl-8-methoxycarbazole (**44**). The latter on treatment with boron tribromide afforded heptazoline (**39**).

(d) OIKAWA and OSAMU (*55*) have synthesised girinimbine (**34**) and murrayacine (**45**), two pyranocarbazole alkaloids, from ketosulphoxide (**47**). Indole (**46**) prepared from (**47**) and (**48**) was cyclised to give dihydrogirinimbine (**49**). (**49**) on treatment with $C_6H_5SO_3Cl$ followed by dehydrogenation and lithium aluminium hydride reduction furnished girinimbine (**34**) which was oxidised to furnish murrayacine (**45**).

III. Biogenesis of Carbazole Alkaloids

Isolation of 3-methylanthraquinone (*24*) from *Clausena heptaphylla* is relevant to the suggestion that mevalonate participates in the formation of the third ring of carbazole alkaloids (*23, 38*). POPLI and KAPIL'S proposal (*38*) that 2-hydroxy-3-methyl-carbazole plays a prominent role in the formation of C_{18}- and C_{23}-carbazole alkaloids has received circumstantial support from the discovery of mukonal (**50**) and 2-hydroxy-3-methyl-carbazole (**7**) in *Murraya koenigii* (*9, 18*). This has also been supported by biomimetic hydroxylation (*63*) of 3-methylcarbazole with FENTON's reagent and under Udenfriend conditions, a reaction prototype of mixed function oxidases. In both systems 2-hydroxy-3-methylcarbazole (**7**) was isolated as the major product in addition to other compounds.

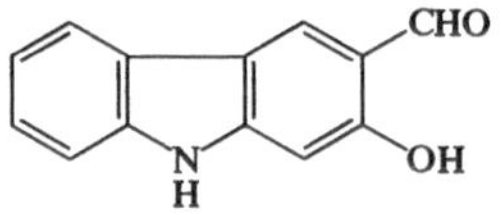

CHO

OH

N
H

(**50**)

Oxidative functional variants of the C-3 methyl group, i.e. CH_2OH, CHO, COOH and COOMe, have also been encountered in various alkaloids, findings which also support the concept of *in vivo* oxidation of carbazole alkaloids (*23*). Recent biosynthetic studies of carbazole alkaloids in lower plants by NAKAMURA *et al.* (*50*) have shown that in Carbazomycin B (*66*) tryptophan contributes to C-3 and C-4 of the hexasubstituted benzene ring in addition to the indole ring. This indicates that tryptophan (**51**) after decarboxylation and deamination reacts with acetate and an unknown C_2-unit before methylation with methionine. However the progenitor of the carbon and methyl substitution at 2-position in Carbazomycin B has not been identified.

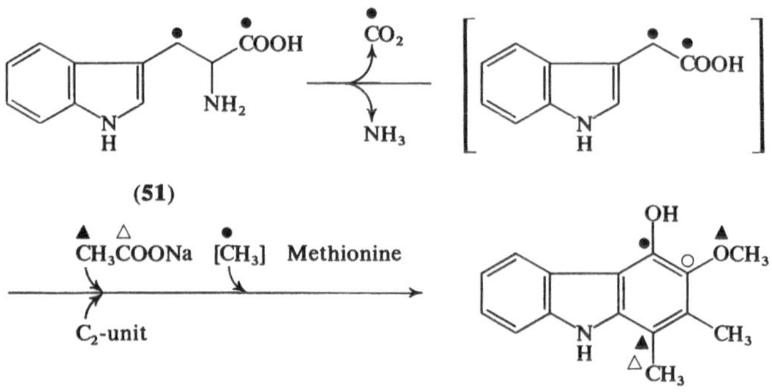

COOH CO_2

NH$_2$ NH$_3$

COOH

N
H

N
H

(**51**)

CH$_3$COONa [CH$_3$]   Methionine

C_2-unit

OH

O   OCH$_3$

CH$_3$

N
H   CH$_3$

IV. Biological Properties of Carbazole Alkaloids and Related Compounds

Carbazole alkaloids are known to exhibit diverse biological activities. A few aspects of this will be discussed. Extensive work in this field is still in progress.

Glycozolidol (52) was found to be active against *Staphylococcus aureus* SH, *Bacillus firmis*, *Sarcina lutea*, *Agrobacterium tumefaciens* and *Proteus vulgaris*.

Carbazomycin A and B (53, 54) are two antibiotics containing a carbazole nucleus. Carbazomycin A exhibited only very weak inhibition of a few fungi and bacteria, while carbazomycin B inhibited the growth of some pathogenic fungi and showed weak antibacterial and antiyeast activities.

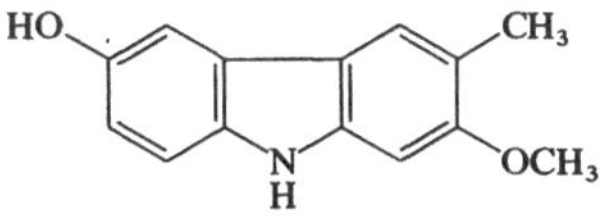

(52) 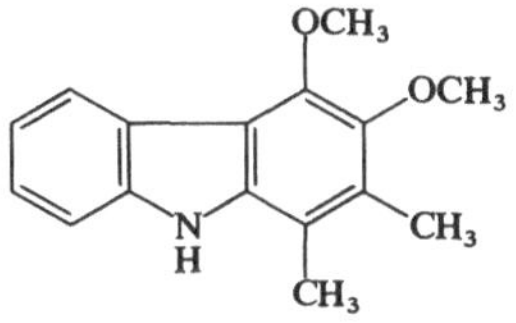　　　　　　(53) 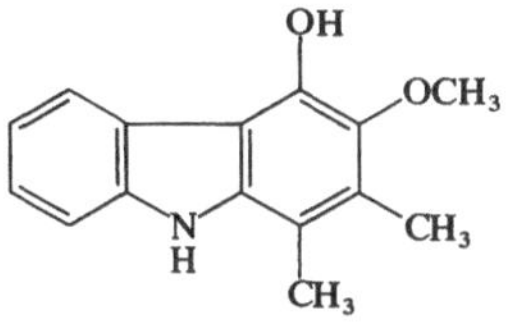　　　　　　(54)

Antifungal and antibacterial properties of some carbazoles isomeric with glycozoline and girinimbine were studied by PATEL *et al.* (60), but the compounds showed little activity. The authors suggested that symmetrical disposition of the methyl and hydroxy groups on the carbazole ring is essential for maximum antifungal activity.

CHOWDHURY *et al.* (29) showed that some tetrahydro carbazoles had low insecticidal properties. Some substituted 1,2,3,4-tetrahydrocarbazoles were found to be active against *Trypanosoma cruzi* (5).

Some acidic tetrahydrocarbazoles have been shown to have antiinflammatory activity (57). Thus 1-Ethyl-8-n-propyl-1 2,3,4,-tetrahydrocarbazole-1-acetic acid (55) is a novel antiinflammatory agent.

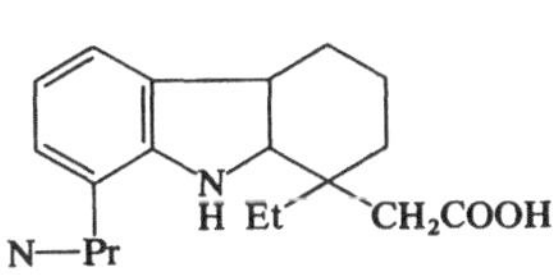

(55) 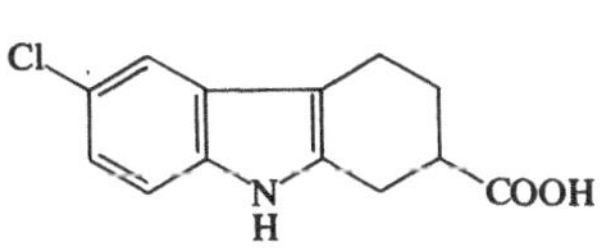　　　　　　(56)

6-Chloro-1,2,3,4-tetrahydrocarbazole-2-carboxylic acid (56) was found to be clinically active in the treatment of acute gout (5).

CARPROFEN (**57**) had remarkable analgesic, antipyretic and antiin-flammatory activity comparable to that of indomethacin (*42*). It had a greater safety margin than indomethacin with respect to the production of gastric ulcers or the blockade of diarrhoea and the reduction of inflammation in adjuvant induced polyarthritis.

(57)   (58)

3-Dimethylamino-1,2,3,4-tetrahydrocarbazole (**58**) which has a modified tryptamine structure prevented amphetamine – induced stereotyped behaviour in rats and prevented reserpine – induced ptosis in mice (*47*). Another potent analogue was 9-ethyl-N,N,1-trimethyl-1,2,3,4 tetrahydrocarbazole-1-ethanamine (**59**).

(59)

N-alkylamino carbazoles (**60, 61**) possess significant anticonvulsant and diuretic activity (*71*). Introduction of aminopropyl chain at the N-atom seems to enhance the anticonvulsant activity in combination with CH_3O at positions 2, 3 and 4.

(60)   (61)

Compound (**62**) was exhibited a growth inhibition at a concentration of 1 µg/ml in mammary carcinoma tissue (*61*)

(62)   (63)

Koenoline (**63**) exhibited cytotoxic activity against the KB cellcul-
ture system (*31*).

(**64**)

Bis-basic ethers of carbazoles are useful as antiviral agents (*3*). 9-Ben-
zoyl-1,2,3,4-tetrahydro-3-hydroxymethylcarbazole (**64**) and its carbani-
late are useful as bactericides and inflammation inhibitors (*4*).

V. Chemistry of Carbazole Alkaloids

A. Members of the C_{13}-Skeleton Group

1. Mukonal

Mukonal, $C_{13}H_9NO_2$ (**50**), m.p. 238° (M^+ 211) was isolated from
the petroleum ether (60–80°) extract of the stem bark of *Murraya koenigii* by BHATTACHARYYA and CHAKRABORTY (*9*). It formed a 2,4-dinitro-
phenylhydrazone and reduced ammoniacal silver nitrate solution indicating the presence of an aldehyde function. The formation of a deep
blue colour with ferric chloride indicated the presence of a chelated
hydroxyl.

The ir spectrum of (**50**) showed it to be an aromatic compound
with -OH or -NH (3380 cm^{-1}) chelated aldehyde (1640 cm^{-1}) while
the uv spectrum (λ_{max} 234, 247, 278, 297 and 342 nm; logε 4.42, 4.21,
4.54, 4.58 and 4.06) was strikingly similar to 3-formylcarbazole suggesting the presence of such a chromophore. The nmr data showed the
presence of a hydroxyl (δ 11.76; 1H, s), an -NH function (δ 11.0; 1H,
brs), one aldehyde (δ 10.16; 1H, s), two aromatic protons one at δ 8.4
(1H, s) and the other at δ 7.0 (1H, s). Four aromatic protons appearing
as a complex multiplet in the region δ 8.0–7.3, suggested that one of
the benzene ring was unsubstituted. The realtively deshielded singlet
at δ 8.4 was assigned to the C-4 proton *ortho* to the aldehyde group.
Moreover as this proton was not *meta* coupled and the hydroxyl was
chelated, the hydroxyl group was placed at C-2. The ^{13}C nmr spectrum
also supported the above assignments. The mass spectrum of mukonal
showed a molecular ion peak at m/z 211 and a high intensity base

peak at M−1 (100%) consistent with the presence of a phenolic group represented by the ionic species (65).

Mukonal, on acetylation formed an acetate (66) m.p. 210° and on decarbonylation formed 2-hydroxycarbazole (5). On oxidation (50) furnished (67) which on partial methylation furnished mukonidine (68) (26). On the basis of these physical and chemical data mukonal was assigned structure (50) which was confirmed by comparison with an authentic specimen of 2-hydroxy-3-formylcarbazole.

2. 2-Methoxy-3-methylcarbazole

2-Methoxy-3-methyl carbazole $C_{14}H_{13}NO$ (69) m.p. 245° was obtained from the petroleum ether (60–80°) extract of the seeds of *Murraya koenigii* (12). The ir spectrum (ν_{max} 3425, 1640, 1600, 1208, 820 and 750 cm^{-1}) and the uv spectrum (λ_{max} 235, 255, 300 and 328 nm; $\log\varepsilon$ 4.35, 3.8, 3.9 and 3.30) indicated it to be a carbazole alkaloid. The nmr spectrum showed signals at δ 8.0 (bs, 1H, NH) 7.5 (s, 1H, H-4) 7.4–7.2 (complex multiplet 4H, aromatic protons) 6.95 (s, 1H, H-1) 3.77 (s, 3H, OCH$_3$) 2.35 (s, 3H, aromatic C-Me).

On zinc dust distillation (69) furnished 3-methylcarbazole (20). Demethylation of (69) with HBr and acetic acid furnished 2-hydroxy-3-methylcarbazole (7). All these evidences led to the formulation of the alkaloid as (69).

CH$_3$

OCH$_3$

N
H

(69)

Zn-dust distillation

Demethylation

CH$_3$

N
H

(20)

CH$_3$

OH

N
H

(7)

3. Mukoline and Mukolidine

Mukoline (70) $C_{14}H_{13}NO_2$ (M$^+$ 227) m.p. 115–20° and mukolidine (71) $C_{14}H_{11}NO_2$ (M$^+$ 225) m.p. 152–55°, two optically inactive carbazole alkaloids, were isolated from the benzene extract of the roots of *Murraya koenigii* (62).

The ir spectrum of mukoline showed the presence of -OH, -NH and an aromatic residue (ν_{max} 3440, 3240 and 1610 cm^{-1}). The uv spectrum (λ_{max} 221, 242, 252, 258, 280, 290 and 320 nm; log ε 4.60, 4.85, 4.65, 4.0, 3.4, 3.6 and 3.0) was strikingly similar to that of 1-methoxycarbazole, thus suggesting the presence of such chromophore in (70). The nmr spectrum showed the presence of an NH (δ 8.22), aromatic protons (δ 7.0–7.9) deshielded benzylic methylene protons (δ 4.75), a hydroxyl proton (δ 4.5) and an aromatic methoxy group (δ 3.9).

On methylation mukoline furnished an N-methyl derivative (72) m.p. 140°. The uv spectrum of (72) was similar to that of 1-methoxycarbazole. The o-acetate of mukoline (73) obtained by acetylation of (70) had an uv spectrum similar to that of (70) indicating that the OH group of mukoline is in the side chain. These observations together with the nmr data suggested the presence of an CH$_2$OH group on the carbazole nucleus of mukoline. This was confirmed by oxidation of (70) to an aldehyde (74) $C_{14}H_{11}NO_2$ m.p. 152–55°, whose uv spectrum was very similar to that of 3-formylcarbazole, but was not identical with that of murrayanine (25). On decarbonylation (74) furnished 1-methoxycarbazole (75). Mukoline was also not identical with 1-methoxy-3-hydroxymethylcarbazole obtained by sodium borohydride reduction of murrayanine. Hence mukoline was formulated as 1-methoxy-6-hydroxymethylcarbazole (70).

The ir spectrum of mukolidine (71) showed the presence of NH
and aldehyde functions in an aromatic system (ν_{max} 3185, 1660 cm^{-1}).
The uv spectrum was characteristic of 3-formylcarbazole. The nmr
spectrum showed the presence of an aldehyde proton (δ 10.8, s, 1H)
an -NH proton (δ 8.6, s, 1H) aromatic protons (δ 8.15, 8.08, 7.3–7.6)
and an aromatic methoxy group (δ 4.05, s, 3H). The nmr data and
mass spectral fragmentation (m/z 225, 210, 182, 167) suggested it was
a 3- or 6-formylcarbazole containing a methoxy group. The position
of the latter was settled by decarbonylation of (71) to 1-methoxycarba-
zole. Mukolidine on borohydride reduction furnished mukoline (70).
All these data led to the formulation of mukolidine as (71). Structures
of mukoline and mukolidine were finally been confirmed by synthesis
as follows.

Condensation of 2-hydroxymethylenecyclohexanome (76) under
Japp-Klingemann conditions with toluenediazonium chloride (77) furnished hydrazone (78) m.p. 205°, which on indolisation furnished the
oxotetrahydrocarbazole (79) m.p. 190°. Dehydrogenation of (79) with
Pd/C at 180° gave 6-methyl-1-hydroxycarbazole (80) which on methylation with diazomethane furnished (81) m.p. 150°. DDQ oxidation of
(81) furnished mukolidine (71). Mukolidine (71) on reduction with sodium borohydride furnished mukoline (70).

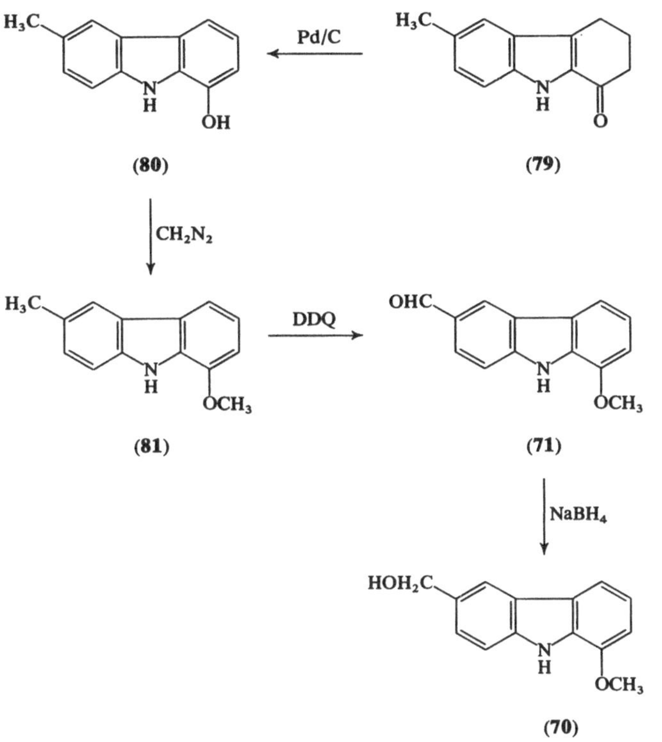

(80)   (79)

(81)   (71)

(70)

4. 2-Hydroxy-3-methylcarbazole

2-Hydroxy-3-methylcarbazole (7) $C_{13}H_{11}NO$ (M^+ 197) m.p. 245° was isolated from the roots of *Murraya koenigii* (*18*). Its ir spectrum showed bands at 3520 (phenolic OH) 3400 (-NH) 1635 and 1600 cm^{-1} (aromatic system). It had a uv spectrum (λ_{max} 235, 254, 258, 304 and 332 nm; log ε 4.65, 4.25, 4.26, 4.19 and 3.66) suggestive of a carbazole skeleton. The nmr spectrum of (7) showed signals at δ 8.2 (bs, 1H, NH) δ 8.1 (s, 1H, OH) δ 8.0–7.1 (complex multiplet, 4H, aromatic proton) δ 7.68 (s, 1H, H–4) δ 7.0 (s, 1H, H-1) δ 2.33 (s, 3H, aromatic methyl).

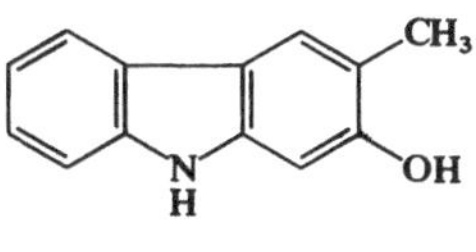

(7)

The 3-methylcarbazole skeleton of (7) was confirmed by the isolation of 3-methylcarbazole on zinc dust distillation of (7). Moreover the acetate of (7) m.p. 210° had a uv spectrum (λ_{max} 236, 262, 296, 330 and 342 nm; log ε 4.40, 4.05, 4.05, 3.30 and 3.30) characteristic of 3-methylcarbazole indicating the presence of a methyl group at the 3-position. Structure assignment (7) was supported by ^{13}C nmr spectroscopy and confirment by direct comparison with an authentic specimen of 2-hydroxy-3-methylcarbazole.

5. Glycozolinol

Glycozolinol (82), $C_{13}H_{11}NO$ (M^+ 197) m.p. 230°, was isolated from the alcohol extract of the root of Glycosmis pentaphylla (19). The compound gave a positive ferric chloride test for phenols. Its ir spectrum showed the presence of an -NH- function (3500 cm^{-1}) phenolic OH (3410 cm^{-1}) C-methyl (1390 cm^{-1}) on an aromatic system (1628, 1570, 1495 cm^{-1}). The uv spectrum (λ_{max} 225, 255, 270, and 302 nm; log ε 4.30, 4.02, 3.9 and 4.2) indicated it to be a carbazole alkaloid. The nmr spectrum showed signals for one OH proton (δ 11.1, s), one NH proton (δ 8.81, s), two aromatic protons (δ 7.8, d) four aromatic protons (δ 7.20–6.86, m) and one aromatic C-methyl (2.5; s) suggesting it to be a carbazole alkaloid with a C-methyl and phenolic hydroxyl group. The mass spectrum of (82) showed the base peak at (M-1) (83) and another significant peak at M-1-28 (84).

On acetylation glycozolinol furnished an acetate (85) m.p. 209° which had a uv spectrum similar to that of 3-methylcarbazole (20). Reduction of the tosylate by Raney Nickel furnished 3-methylcarbazole (20). On methylation with diazomethane, glycozolinol furnished glycozoline (86). The formulation of glycozolinol as (80) has been confirmed by synthesis as follows.

Condensation of 2-hydroxy-methylene-5-methylcyclohexanone (87) with diazotised 4-hydroxybenzene diazonium chloride (88) under Japp-Klingemann reaction condition furnished 4-methylcyclohexane-1, 2-dione-1-(4'-hydroxy)-phenylhydrazone (89). (89) on indolisation gave the oxotetrahydrocarbazole (90). This on reduction and subsequent dehydrogenation furnished (91) and (82) respectively.

(83)   (84)

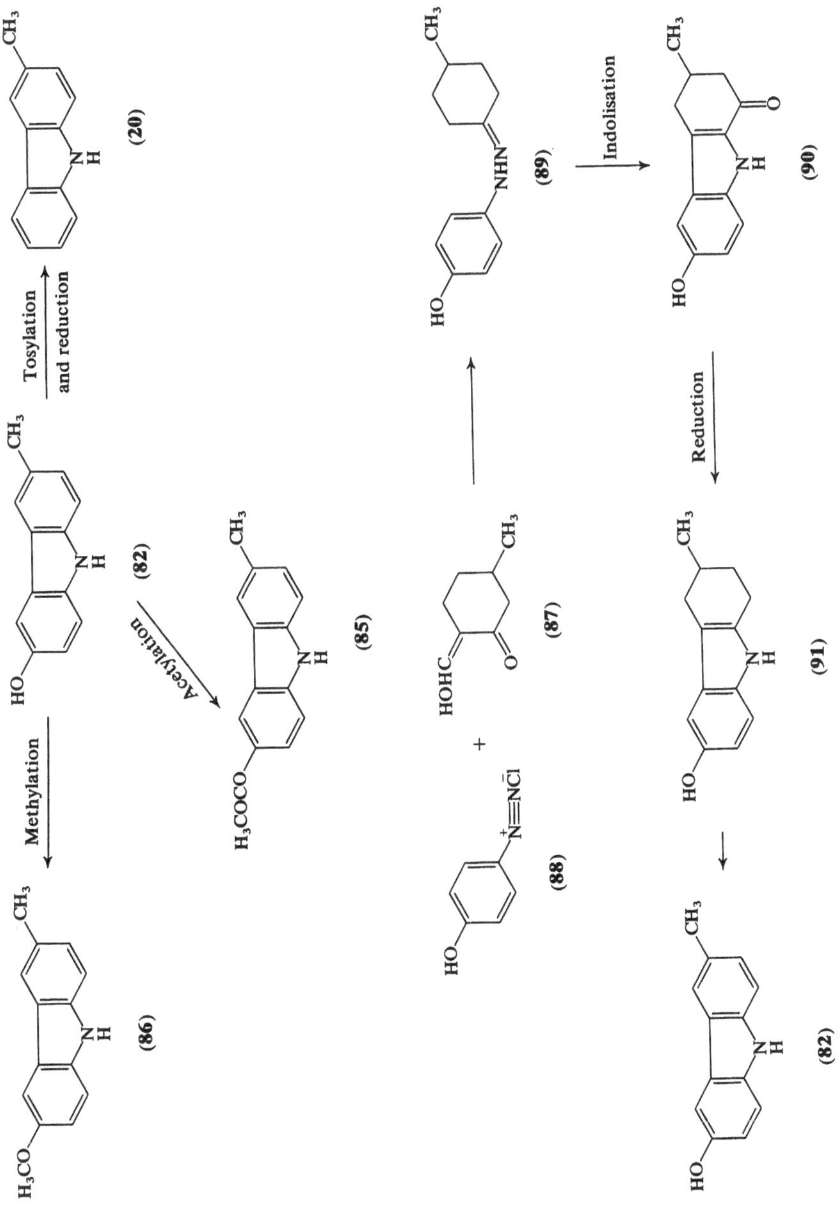
Tosylation and reduction
Methylation
Acetylation
Indolisation
Reduction
(20)
(82)
(85)
(86)
(87)
(88)
(89)
(90)
(91)
(82)

Mukherjee *et al.* (*49*) have isolated the same alkaloid from the seeds of *Glycosmis pentaphylla* and named it Glycozolinine.

6. Glycozolidol

Glycozolidol (**52**), $C_{14}H_{13}NO_2$ (M^+ 227) m.p. 240° was isolated from the benzene extract of the root of *Glycosmis pentaphylla* (*15*).

Glycozolidol has one hydroxyl group indicated by green colouraction with $FeCl_3$. The ir spectrum showed peaks at v_{max}^{KBr} 3500 (-OH) 3440 (-NH) 1625, 1600 (aromatic residue) 1380 (C-Me) 1208 (aromatic ether) and 815 cm^{-1} (substituted benzene derivative). The uv spectrum of (**52**) (λ_{max} 232, 260, 305 nm; log ε 4.50, 4.10 and 4.29) was suggestive of the presence of a carbazole system. The nmr spectrum of (**52**) showed signals for one hydroxyl function as a singlet at δ 10.8 and an -NH as a broad singlet at δ 8.0. The C-1 and C-4 protons appeared as singlets at δ 6.92 and 7.45 respectively. The C-5 and C-8 proton appeared as doublets at δ 7.2 and 7.08. The C-7 proton appeared as a double doublet at δ 6.75, the methoxyl proton as a singlet at δ 3.7 and the aromatic C-Me as a singlet at δ 2.38. The appearance of H-4 as a singlet suggested substitution at positions 3 and 2 and the lack of ortho coupling for H-5 suggested substitution at position 6 by the

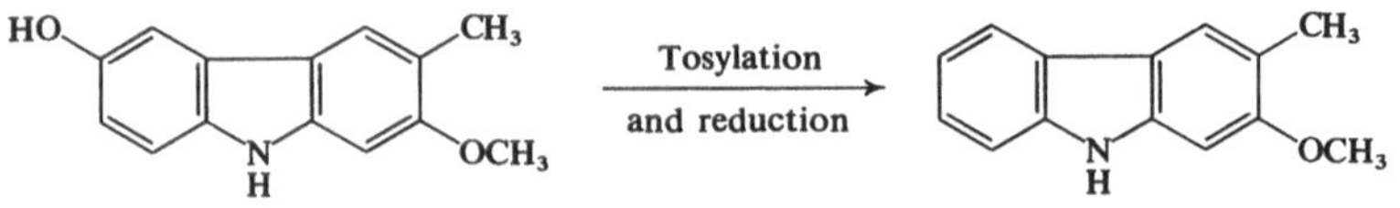

(52)   (69)

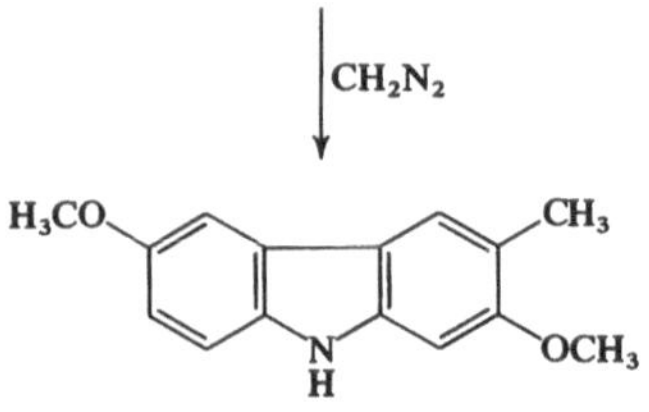

(94)

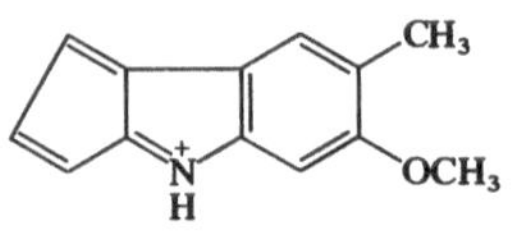

(93) 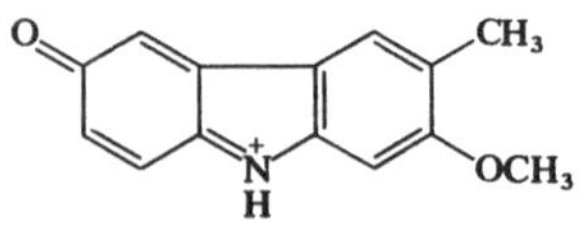   (92)

hydroxyl to C-6. The mass spectrum of (52) showed a molecular ion peak at m/z 227. The base peak at m/z 226 and an other significant peak can be represented by ionic species, (92) and (93).

The acetate of glycozolidol had uv spectrum similar to that of 2-methoxycarbazole suggesting the presence of the methoxy on position 2. This was verified by reduction with Raney nickel of the tosyl derivative of glycozolidol to 2-methoxy-3-methylcarbazole (69) which showed the presence of a 2-methoxy-3-methylcarbazole skeleton and led to the formulation of glycozolidol as (52). The structure was confirmed by methylation of glycozolidol with diazomethane to glycozolidine (94).

7. Glycozolidal

Glycozolidal (95) $C_{15}H_{13}NO_3$ (M$^+$ 255) m.p. 185° was isolated from the petroleum ether (40–60°) extract of the root of *Glycosmis pentaphylla* (*13*). The ir spectrum of (95) showed it to be an aromatic compound with -NH- (3500 cm^{-1}), aldehyde (1675 cm^{-1}), aromatic ether (1220, 1200 cm^{-1}) while the uv spectrum was strikingly similar to that of 3-formylcarbazole (λ_{max} 235, 248, 303, 340 nm; log ε 4.4, 4.2, 4.2, 4.06). The nmr data showed the presence of an -NH- function (δ 8.8, brs, 1 H) one aldehyde group (δ 9.9, s, 1 H) two aromatic methoxyl (δ 3.9, s, 3 H; 3.7, s, 3 H) and four aromatic protons at (8.4, s, 1 H, H-4) (δ 7.6, d, 1 H, H-5) (δ 6.85 dd, 1 H, H-7) (δ 7.3, d, 1 H, H-8). All these data suggested presence of a carbazole system with two aromatic methoxyls and an aldehyde function and led to the formulation of glycozolidal as (95). The structure was confirmed by oxidation of glycozolidine (94) with DDQ to a compound identical with natural glycozolidal.

(94)   (95)

8. Lansine

Lansine (37) $C_{14}H_{11}NO_3$ (M$^+$ 241) m.p. 225–26° was isolated from the ethanolic extract of the leaves of *Clausena lansium* (Lour.) Skeels (syn. *C. wampi* Oliv.) (*58*). Its ir spectrum showed the presence of an NH and chelated carbonyl on an aromatic system (ν_{max}^{KBr} 3300, 1645 and 1620 cm^{-1}). The uv spectrum (λ_{max} 233, 248, 309, 340 and 368 nm) was characteristic of 3-formylcarbazole which shifted to (λ_{max} 245, 312 and 380 nm) on addition of alkali. The nmr spectrum showed signals for one methoxyl (δ 3.86, s, 3H) one aldehyde (δ 9.91, s, 1H) H-1 and H-4 protons (δ 6.80, s, 1H; δ 8.36, s, 1H), H-5 proton (δ 7.61, d, 1H), H-7 proton (δ 6.91, dd, 1H) and H-8 proton (δ 7.30, d, 1H). From these evidences the structure of lansine was formulated as (37) and was confirmed by conversion of lansine to the known 2,6-dimethoxy-3-formylcarbazole (95) on refluxing with methyl iodide. The structure was further confirmed by synthesis as follows.

DDQ oxidation of 2,6-dimethoxy-3-methylcarbazole (94) furnished 2,6-dimethoxy-3-formylcarbazole (95) m.p. 185°, whose selective demethylation with BCl$_3$ in dichloromethane furnished lansine (37).

9. Koenoline

Koenoline, (63) $C_{14}H_{13}NO_2$, m.p. 213° (M$^+$ 227), was isolated from the chloroform extract of the root bark of *Murraya koenigii* (31). The ir spectrum of (63) showed it to be an aromatic compound with -NH- (3445 cm^{-1}) and -OH (3235 cm^{-1}) while the uv spectrum (λ_{max} 335, 323, 289, 279, 258, 251, 241 and 225 nm; log ε 3.58, 3.61, 4.04, 3.86, 4.38, 4.22, 4.71, and 4.56) was strikingly similar to that of a 1-methoxycarbazole derivative (23). The non-phenolic nature of the hydroxyl group was suggested from its negative colour reaction with FeCl$_3$. On acetylation (63) furnished a monoacetate (96) m.p. 110°. The nmr spectrum of the alkaloid showed signals of an OH proton (δ 1.75, brs), NH proton (δ 8.34, brs), H-2 (δ 6.95, brs), H-4 (δ 7.66, brs), H-5 (δ 8.04, d, J=8.0 Hz), H-6 (δ 7.23, ddd, J=8.0, 7.0, 1.3 Hz) H-7 (δ 7.42, ddd, J=8.0, 7.0, 1.2 Hz), H-8 (δ 7.46, d, J=

8.0 Hz), an aromatic methoxyl (δ 4.01, s) and a benzylic methylene (δ 4.84, s). As the H-6 and H-7 protons both showed one *meta* and two *ortho* couplings it was suggested that one of the benzene ring of the carbazole nucleus was unsubstituted, although the signals attributed to H-8 and H-5 showed only *ortho* coupling. The H-4 and H-2 protons were singlets suggesting substitution at C-1 and C-3. The above assignments were substantiated by decoupling experiments and by the ^{13}C nmr spectrum. Finally the structure of koenoline was confirmed as (63) by its partial synthesis from murrayanine (97). The latter on sodium borohydride reduction furnished a compound which was identical with (63) in all respects.

(63)   acetylation →   (96)

(97)   NaBH$_4$ →   (63)

10. Murrayafoline-A

Murrayafoline-A (98), $C_{14}H_{13}NO$, m.p. 52–53° (M^+ 211), was obtained from the ethanolic extract of the root bark of *Murraya euchrestifolia* Hayata (33) as colourless plates. The ir spectrum of the compound (ν_{max} 3480, 1640, 1610, 1590, 1505 cm^{-1}) showed the presence of an imino group on an aromatic system. The uv spectrum (λ_{max} 225, 243, 251 sh, 283 sh, 292, 330 and 344 nm; log ε 4.47, 4.58, 4.44, 3.83, 4.01, 3.53 and 3.49) was characteristic of 1-methoxycarbazole (23). The nmr spectrum showed signals of an NH proton (δ 7.96), H-2 (δ 6.55, s) H-4 (δ 7.33, s) H-5 (δ 7.87, d, J = 8 Hz) H-6, H-7 and H-8 (δ 6.9–7.3, m). In addition to these an aromatic C-methyl singlet at δ 2.42 and a methoxyl singlet at δ 3.76 showed the presence of methyl and methoxyl substitution on the carbazole nucleus. The multiplet for three protons at δ 6.9–7.3 along with the H-5 signal as doublet at δ 7.87 suggested that one of the benzene rings of the carbazole nucleus was unsubstituted.

The atachment of the methyl and methoxyl groups to the C-3 and C-2 respectively was deduced from experiments involving nuclear Overhauser effects. Observation of a 18.5% NOE enhancement between H-2 at δ 6.55 and methoxyl at δ 3.76 and observations of 9.6% and 9.5% enhancements between H-4, H-2 and the aryl methyl at δ 2.42 located the methoxyl on C-2 and the methyl on C-3, thus leading to (98) as the structure of murrayafoline.

(98)

B. Members of the C_{18}-Skeleton Group

1. Mukonicine

Mukonicine (99), $C_{20}H_{21}NO_3$, (M$^+$ 323), m.p. 233–34°, was isolated from the alcoholic extract of the leaves of *Murraya koenigii* (48). The ir spectrum of mukonicine showed bands for an -NH- function, -OMe, an aromatic C-Me, an aromatic residue and a substituted benzene derivative (v_{max} 3440, 1648, 1630, 1560, 1460, 1385, 765 and 745 cm^{-1}). Its uv spectrum (λ_{max} 226, 240, 300 and 342 nm; log ε 4.70, 4.67, 4.59 and 4.26) was very similar to koenimbine (52) showing the presence of the same pyranocarbazole chromophore in the two alkaloids. The nmr signals of mukonicine showed the presence of an -NH proton (δ 7.8, s), two olefinic protons (δ 6.55, d; δ 5.65, d), six protons of two aromatic methoxyls (δ 3.9, s) and three protons of an aromatic C-Me (δ 2.3, s). The C-4 proton appeared as a singlet at δ 7.56 which suggested that positions 2 and 3 were substituted. Like koenimbine, the C-5 proton appeared at higher field (δ 7.40, s) thus suggesting that one of the methoxyl was at C-6. The second methoxyl was placed at C-8 as the C-7 proton appeared at a much higher field (δ 6.88, s) than that of C-5. A six proton singlet at 1.44 togeher with symmetrical doublets at δ 6.55 and 5.65 (J = 9 Hz each) suggested the presence of a 2,2-dimethyl-Δ^3-pyran ring fused to the carbazole system. The mass spectrum of mukonicine showed a molecular ion peak at m/z 308 (M-15) which supports the presence of a 2,2-dimethyl-Δ^3-pyran system in mukonicine and could be represented by the carbazolopyrilium ion (100). Further proof for the presence of a 2,2-dimethyl-Δ^3-pyran ring

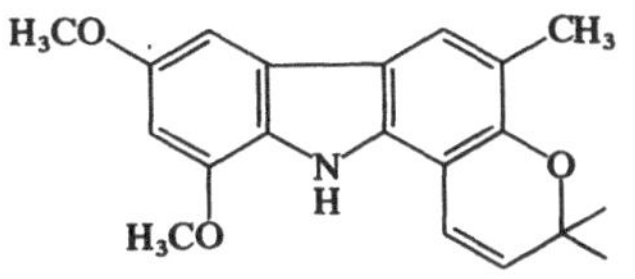

(99)

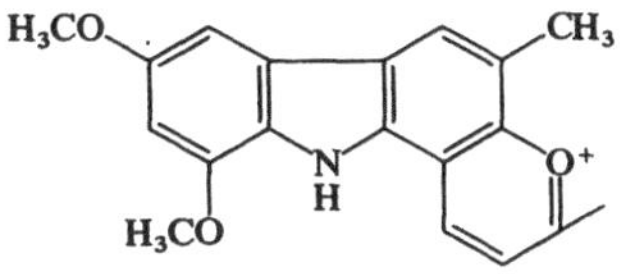

(100)

was provided by chromic acid oxidation of mukonicine when acetone was obtained.

The 3-methylcarbazole skeleton of mukonicine was confirmed by isolation of 3-methylcarbazole on zinc dust distillation of mukonicine.

From all these facts and the non-identity of mukonicine with koenigicine (51) structure (99) was assigned to mukonicine.

2. Heptazolicine

Heptazolicine (101), $C_{18}H_{17}NO_3$ (M$^+$ 295), m.p. 285° (dec), was isolated from the alcoholic extract of the roots of *Clausena heptaphylla* (11) and was shown to have phenolic and aldehyde functions. Its ir spectrum showed absorption peaks at 3260 (OH, hydrogen bonded) 3000 (NH) 1700 (CHO) 1615 and 1517 cm^{-1} (aromatic) while the uv spectrum (λ_{max} 242, 275 and 300 nm log ε 4.65, 4.62 and 4.40) indicated the presence of a 3-formylcarbazole chromophore. Its nmr spectrum showed that the signal for C-4 proton was deshielded (δ 8.3, s) due to a vicinal aldehyde function. The proton at C-5 was *ortho* and *meta* coupled (δ 7.5, d, J = 8 Hz and 3 Hz) showing that positions 6 and 7 were not substituted. The protons at C-6 and C-7 appeared as a complex multiplet (δ 6.7–7.2). Two symmetrical triplets (δ 3.03 and 1.98) and a sharp six proton singlet at δ 1.45 account for the 2,2-dimethyldihydropyran ring. The presence of 2,2-dimethylpyran system was also supported on mass spectral evidence. The mass spectrum showed molecular ion peak at m/z 295. The other significant peak at m/z 280 represented the ionic species (102).

On acetylation, heptazolicine furnished an acetate (103) m.p. 230°, whose uv spectrum was very similar to that of cycloheptaphylline

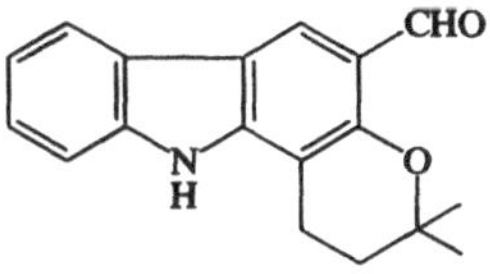

(103 A)

(101)   (103)

(102)

(103 A) (35) suggesting the presence of the same chromophore. Finally heptazolicine was found to be identical in all respects with the cyclization product of heptazoline (23). Hence the structure of hepatazolicine is (101).

3. Clausenapin

Clausenapin (104), $C_{19}H_{21}NO$ (M^+ 279), m.p. 101°, was isolated from the leaves of Clausena heptaphylla (14). The ir spectrum showed bands at 3350 (-NH), 1600, 1570 (aromatic residue) 1375 (C-CH$_3$), 1208 (aromatic ether) 765 and 740 cm^{-1} (substituted benzene derivative). The uv spectrum (λ_{max} 224, 240, 250, 290, 320 and 335 nm; log ε 4.55, 4.59, 4.50, 4.29, 5.39 and 5.50) was very similar to that of 1-methoxycarbazole. The nmr signals of clausenapin showed the presence of an -NH function (δ 8.8, s), a proton at C-4 (δ 7.5, s), a proton at C-5 (δ 7.9, q, J = 7.0 Hz and 2.0 Hz), three aromatic protons as a multiplet at δ 7.4–7.1, one aromatic c-methyl (δ 2.28, s) besides a vinylic proton (δ 5.3, t), two protons of a benzylic methylene (δ 3.62, d) and 3 protons each of two methyl groups (δ 1.85 and 1.75) which are parts of an isopentenyl chain.

Zinc dust distillation of clausenapin led to the isolation of carbazole and 3-methylcarbazole, indicating the presence of 3-methylcarbazole residue in the molecule. In consideration of the singlet nature of the C-4 proton and the substitution at the 1- and 3-positions, the isopentenyl chain was located in the 2-position. On heating clausenapin with HBr/acetic acid, a cyclised product (105) $C_{18}H_{19}NO$ (M^+ 265) m.p. 135° was obtained. Its nmr spectrum indicated the presence of a 2,2-dimethyldihydropyran ring in (105).

From this evidence, clausenapin has been assigned structure (104).

(104)   (105)

4. Clausenatin

Clausenatin (106), $C_{18}H_{17}NO_2$ (M$^+$ 279), m.p. 154–56°, was isolated from the hexane extract of the roots of *Clausena anisata* (Willd) Oliv. (56). Its uv spectrum had (λ_{max} 239, 249, 278, 288, 297 and 340 nm; log ε 4.40, 4.29, 4.39, 4.46, 4.53 and 3.90); ir bands appeared at 3400 (-OH), 3300 (-NH) 1635 (weak chelated aldehyde) and 1610 cm^{-1} (aromatic) and its colour reactions showed it to be a 3-formylcarbazole derivative with a chelated aldehyde and phenolic hydroxyl group. The nmr spectrum showed the presence of a chelated hydroxyl (δ 11.75, s) and an aldehyde group (δ 9.87, s) and an NH signal at δ 7.85. The aromatic protons at C-4, C-8 and C-1 appeared each as singlet at δ 8.0, 7.37 and 7.34 respectively, while the C-5 proton was a doublet (δ 7.9, J = 7 Hz) and the C-6 proton was quartet (δ 7.31; J = 7, 2 Hz). The two methyl signals (δ 1.72 and 1.85), a benzylic methylene doublet at δ 3.62 (J = 7 Hz) together with the signal for a vinylic proton at δ 5.33 could account for a γ,γ-dimethylallyl residue in clausenatin.

On the basis of the nmr data and on biogenetic grounds the OH was placed at the 2-position. The γ,γ-dimethylallyl group was placed at C-7 because of the nmr spectrum and because the corresponding 6-derivative was found to be different from clausenatin.

(106)

5. Mupamine

Mupamine, $C_{19}H_{19}NO_2$ (107), m.p. 151–52° (M$^+$ 293), was isolated from the aqueous methanolic extract of the root bark of *Clausena anisata* (Willd) Oliv. (45). The uv and ir spectra were suggestive of the presence of a pyranocarbazole system like that of girinimbine. The nmr spectrum of the alkaloid showed signals of an NH proton (δ 8.03), C-4 (δ 7.63), C-5 (δ 7.53) and two other aromatic protons at δ 7.09 and 6.80. A six proton singlet at δ 1.48 together with doublets at 6.65 and 5.68 (J = 9.8 Hz each) for two vinylic protons showed the presence of a 2,2-dimehyl-Δ^3-pyran ring. In addition to these an aromatic C-methyl singlet at δ 2.33 and a methoxyl singlet at δ 3.99 showed the presence of methyl and methoxyl substitution on the carbazole nucleus. The C-4 proton was a singlet suggesting that positions 2 and 3 were substituted. The proton at C-5 was *ortho* and *meta* coupled showing that positions 6 and 7 were not substituted. The angular orientation of the 2,2-dimethyl-Δ^3-pyran ring was deduced from the downfield shift (about 0.52 ppm) of the 4′-proton of the chromene ring in deuterated DMSO compared with that in $CHCl_3$. The mass spectrum of (107) showed a molecular ion peak at m/z 293. The base peak at m/z 278 (M-15) also supported the presence of a 2,2-dimethylpyran system in (107) and can be represented by (108).

From these data the structure of mupamine was assigned as (107) which has also been supported by ^{13}C nmr spectrometry.

Finally the assigned structure of mupamine has been confirmed by two syntheses.

In the synthesis of Reisch *et al.* (*46*) condensation of 2-hydroxy-methylenecyclohexanone (76) with 3-hydroxy-4-methylbenzene diazonium chloride (109) gave (110), which on indolisation gave the oxotetrahydrocarbazole (111). (111) on acetylation gave (112) which on dehydrogenation with Pd/C furnished (113). (113) on methylation with diazomethane furnished 2-acetoxy-8-methoxy-3-methylcarbazole (114). (114) on boiling with 2N NaOH gave 2-hydroxy-8-methoxy-3-methyl-carbazole (115), which on treatment with 3-chloro-3-methyl-1-butyne (35) furnished mupamine (107).

In the synthesis of Sharma *et al.* (*70*) condensation of 2-hydroxy-8-methoxy-3-methylcarbazole (115) with 3-hydroxyisovaleraldehyde dimethylacetal (116) furnished mupamine (107).

(108)

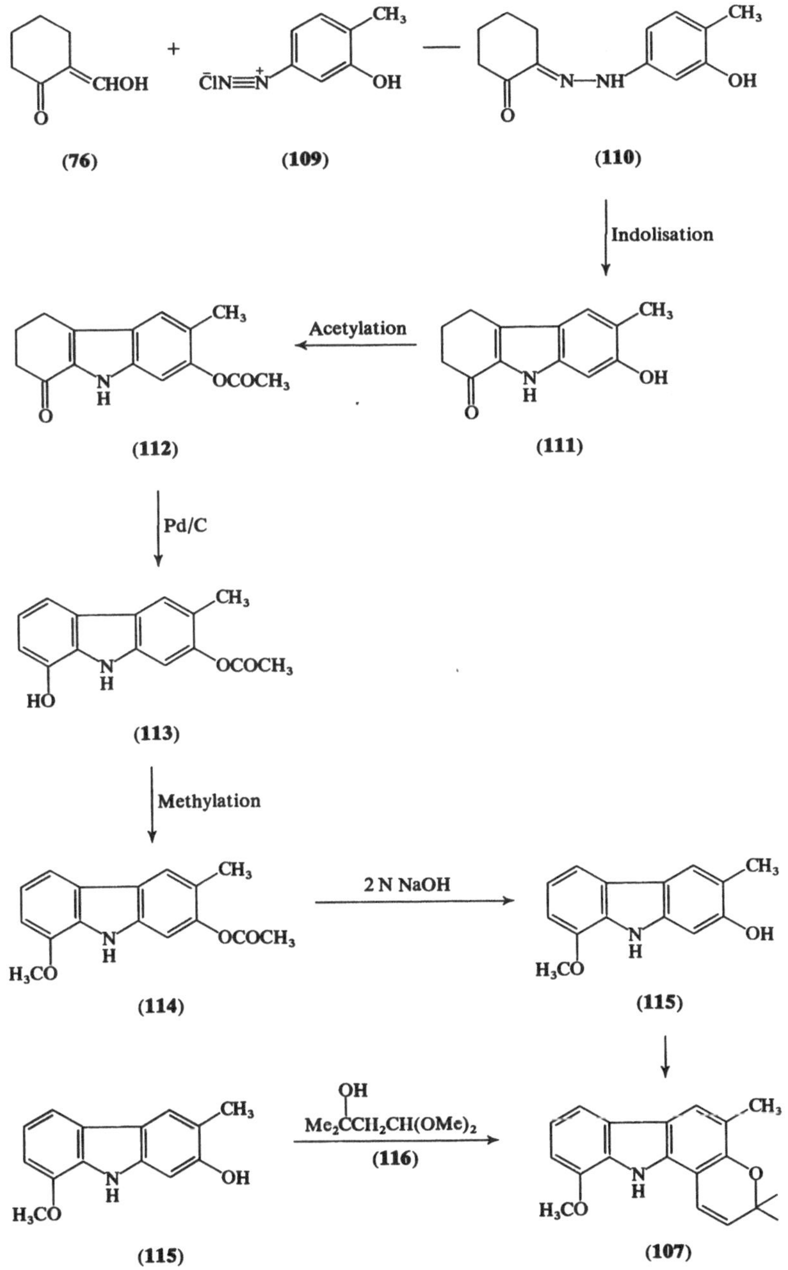

(76)　　　　　　(109)　　　　　　　　(110)

Indolisation

(112) ← Acetylation ← (111)

Pd/C

(113)

Methylation

(114) → 2 N NaOH → (115)

(115) → Me$_2$CCH$_2$CH(OMe)$_2$ (116) → (107)

6. Murrayafoline-B

Murrayafoline-B, $C_{19}H_{21}NO_2$ (117), a colourless syrup (M$^+$ 295), was isolated from the root bark of *Murraya euchrestifolia* (33). The ir spectrum (ν_{max} 3600, 3475, 1620 and 1595 cm^{-1}) and the uv spectrum (λ_{max} 230 sh, 247, 255 sh, 302, 324 sh and 337 sh nm) were suggestive of a carbazole skeleton. The uv spectrum showed a bathochromic shift on addition of NaOH (λ_{max} 234, 254, 304 and 338 nm) indicating the presence of a phenolic hydroxy group. In the nmr spectrum of murraya-foline-B (117) the H-2 and H-4 protons appeared as singlets at δ 6.56 and 7.32. An AB system at δ 7.72 and 6.82 (each 1 H, d, J = 9 Hz) was assigned to the H-5 and H-6 protons respectively. Two one proton singlets at δ 7.94 and 4.99 were assigned to an NH and hydroxy proton respectively. A signal of an aromatic methoxyl group appeared at δ 3.90. Signals for two methyl groups on a double bond (δ 1.88, 1.72) together with signals at δ 5.3 for a vinylic proton and two benzylic protons at δ 3.6 accounted for a γ,γ-dimethylallyl group on an aromatic system.

Photooxidation of murrayafoline-B (117) in methanol furnished a compound, m.p. 223°, which was identical with murrayaquinone-B (118) in all respects (ir, nmr, m.m.p., t.l.c.) thus leading the structure (117) for murrayafoline-B.

(117)   (118)

7. Ekeberginine

Ekeberginine (119), $C_{19}H_{19}NO_2$ (M$^+$ 293), m.p. 230–31°, was isolated from the stem bark of *Ekbergia senegalensis* (Meliaceae) (40). The substance readily formed an N-methyl derivative (120), m.p. 155–57°.

The nmr spectrum of (120) showed signals for an aldehyde (δ 10.37) methoxyl (δ 3.99), C-5 (δ 8.10), C-6 (δ 7.30), C-7 (δ 7.51) C-8 (δ 7.41) C-2 (δ 7.43) and N-methyl (δ 4.13). Two methyl signals (δ 1.70, 1.89), a benzylic methylene (δ 4.19), and vinylic proton (δ 5.28) accounted for a a γ,γ-dimethylallyl residue. The nmr data are closely related to those of indizoline and heptaphylline (23). The ^{13}C nmr spectrum of N-methylekberginine was consistent with the carbazole ring system and the above substituents. The direct assignment of all the protonated carbons of N-methylekeberginine was made by 2D one- bond δC/δH correlation spectrometry which resulted determination of the structure of N-methyl ekeberginine as (120) and hence the structure of ekebergin-ine as (119).

(119)   (120)

C. Members of the C$_{23}$-Skeleton Group

1. Exozoline

Exozoline (121), C$_{23}$H$_{27}$NO (M$^+$ 333), m.p. 180–82°, was isolated from the alcoholic extract of the leaves of *Murraya exotica* (34). Its uv spectrum (λ$_{max}$ 242, 256 and 306 nm; log ε 4.47, 4.18 and 3.97) was characteristic of a 2-methoxycarbazole chromophore. The ir spectrum showed the presence of an -NH function and substituted aromatic system (ν$_{max}$ 3440, 1625, 1615, 1445, 1380, 780 and 740 cm^{-1}). The nmr spectrum was very similar to that of cyclomahanimbine (124) (23) except that a doublet for three protons appeared at δ 0.62 instead of two methylene protons at δ 4.72. The mass spectrum showed the molec-ular ion peak at m/z 333 and two other significant peaks at m/z 248 and 210 representing the species (122) and (123).

(121)   (122)

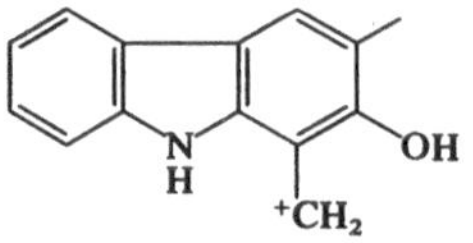

(123)

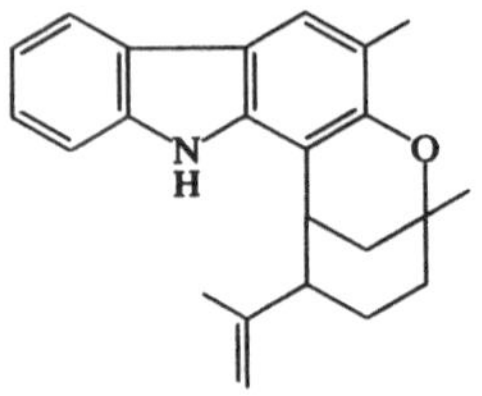

(124)

From this evidence exozoline was formulated as dihydrocycloma-hanimbine (121). Its identity was further established by direct comparison with dihydrocyclomahanimbine obtained by hydrogenation of cyclomahanimbine (124).

2. Mahanimbinol

Mahanimbinol (125), $C_{23}H_{27}NO$ (M^+ 333), was obtained as an amorphous powder from the hexane extract of the stem wood of *Murraya koenigii* (59). Its uv spectrum (λ_{max} 244, 260, 296, 325 and 339 nm; log ε 4.26, 4.14, 3.98, 3.40 and 3.38) was strikingly similar to that of 2-hydroxy-3-methylcarbazole indicating the presence of this chromophore. The ir spectrum showed an absorption band at 3400 cm^{-1} (-NH or OH or both). The nmr spectrum showed the presence of C-4 and C-5 protons at δ 7.50 and 7.83 respectively. Of these C-4 proton was a singlet suggesting that positions 2 and 3 were substituted. The C-5 proton was a multiplet. The C-6, C-7 and C-8 protons appeared as a complex multiplet in the region δ 6.71–7.23. A three proton singlet at δ 2.30 accounted for the aromatic methyl group. The spectrum also showed the presence of a C-geranyl group attached to an aromatic nucleus (three methyl singlets at δ 1.58, 1.73 and 1.83, two allylic methylene groups represented by a four proton signal at δ 2.05, a two proton

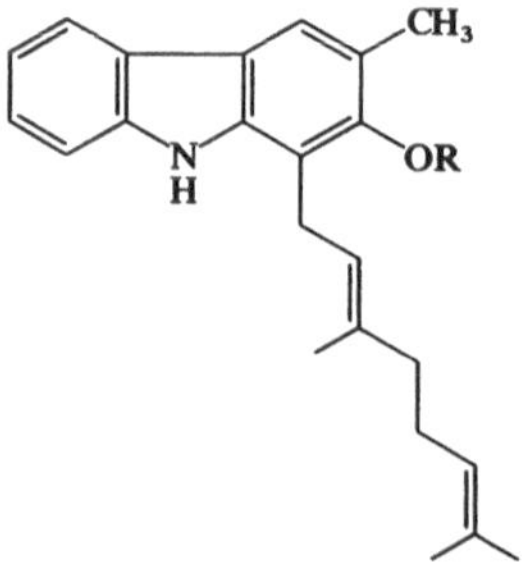

(125) R=H
(126) R=OAC

doublet at δ 3.43 of benzylic hydrogens and a broad triplet of two olefinic protons at δ 5.0 and 5.39). The mass spectrum showed a base peak at m/z 210 formed by the loss of a C_9H_{15} residue from the C_{10} side chain. A strong peak at m/z 209 also indicated the presence of an OH adjacent to the geranyl group.

On acetylation (125) formed a monoacetate (126) m.p. 69° and on hydrogenation a tetrahydroderivative, m.p. 115°.

From this, the structure of mahanimbinol was deduced as (125).

3. Mahanimboline

Mahanimboline (127), $C_{23}H_{25}NO_2$ (M^+ 347), m.p. 170–72°, was isolated from the petroleum ether (60–80°) extract of the root bark of *Murraya koenigii* (65).

The ir spectrum of mahanimboline showed bands for an -OH function, an -NH and an aromatic residue (v_{max} 3440, 3300, 1650 and 1615 cm^{-1}). The uv absorption spectrum (λ_{max} 237, 280, 288 and 330 nm; log ε 4.42, 3.69, 4.25 and 2.75) was very similar to pyranocarbazole alkaloids like girinimbine and mahanimbine and thus indicated the presence of the same pyranocarbazole chromophore in (127). Presence of the 3-methylcarbazole skeleton was established by isolation of 3-methylcarbazole on zinc dust distillation of mahanimboline.

The mass spectrum of mahanimboline showed a molecular ion peak at m/z 347, and a peak at m/z 329 due to loss of H_2O. A high intensity peak at m/z 248 could represent the same carbazolopyrilium ion (122) as that formed from girinimbine and mahanimbine. The nmr spectrum showed signals at δ 8.7 (s, 1 H) for an NH proton, five aromatic protons at δ 8.1 (d, 1 H) 7.8 (s, 1 H) 7.5–7.25 (m, 3 H) and olefinic protons at C-4' and C-5' at 7.0 and 5.7 (d, 2 H). The environment of the aromatic protons was very similar to that of mahanimbine suggesting the same pattern of substitution in mahanimboline. The other signals were

(127) R=CH₂CH₂CH—C

(128) R=CH₂CH₂CH—C

at δ 6.0–5.8 (m, 3H due to two terminal methylene protons and one overlapping CHOH proton) and four protons at δ 2.6 (br, 4H, methylene protons, deshielded by oxygen of the hydroxyl group). From the nmr and mass spectral data it was inferred that the $C_6H_{11}O$ residue contained a secondary hydroxyl group and a terminal methylene group.

On acetylation mahanimboline furnished an acetate (128) m.p. 160°. This suggested the secondary nature of the hydroxyl function. From all these facts mahanimboline has been given the structure (127).

4. Isomurrayazoline

Isomurrayazoline (129), $C_{23}H_{25}NO$ (M$^+$ 331), m.p. 269–70°, $[\alpha]_D^{CHCl_3} - 7.33°$, was isolated from the benzene extract of the stem bark of *Murrraya koenigii* (8). Its uv spectrum (λ_{max} 240, 253, 258, 303 and 330 nm; log ε 4.65, 4.15, 4.08, 3.85 and 3.47) was suggestive of a 2-oxygenated carbazole chromophore. The ir spectrum (λ_{max} 1605, 1380, 1375, 888 and 810 cm^{-1}) lacked an NH peak. The nmr spectrum showed signals for five aromatic protons [δ 7.95 (1H, m); 7.55 (2H, unresolved doublet) 7.13–7.33 (2H, m)]. The environment of the aromatic protons was very similar to that of isomahanimbine (36). Other signals were at δ 3.3 (1H, m, benzylic), δ 2.35 (3H, s, one aliphatic methyl group) and δ 1.47 (7H, m; six methylene proton plus the cyclohexane ring methane proton). From the nmr data it was evident that like isomahanimbine, isomurrayazoline was a 6-methylcarbazole with a fused monoterpene system like murrayazoline (20). The mass spectrum of isomurrayazoline was very similar to murrayazoline.

Isomurrayazoline on acid catalysed hydration furnished isomurrayazoline (130) m.p. 159°. The ir spectrum of (130) showed bands for OH and NH (ν_{max} 3550 and 3290 cm^{-1}). The uv spectrum (λ_{max} 240, 253, 258, 304 and 330 nm; log ε 4.60, 4.12, 4.07, 3.87 and 3.52) was similar to that of a 2-oxygenated carbazole chromophore.

The nmr spectrum of (130) showed signals for an NH proton at δ 8.0 (1H, m) and an OH signal at δ 1.68 (1H, s, exchangeable with D$_2$O). The environments of the aromatic protons were similar to those of (129). Other signals were at δ 3.83 (1H, m, benzylic) δ 2.35 (3H, s, aromatic C-methyl) δ 1.48 (7H, m, six methylene protons plus the cyclohexane ring methane) δ 1.31 (3H, s, alipathic methyl). The two methyls of the gem-dimethyl groups appeared as a singlet at δ 1.90 and 0.53. In comparison with (129), the signal of the benzylic protons of (130) experienced a downfield shift (δ 0.53) due to hydration, a postulate also supported by the mass spectrum of (130).

From these observations and from the close resemblance of the

(129)   (130)

physical properties of isomurrayazoline and isomurrayazolinine to those of murrayazoline and murrayazolinine, structure (129) has been assigned to isomurrayazoline.

5. (+) Murrayazoline

(+) Murrayazoline (131) $C_{23}H_{25}NO$ (M^+ 331), $[\alpha]_D^{CHCl_3} + 2.25$, m.p. 276–278°, was isolated as colourless needles from the root bark of *Murraya euchrestifolia* Hayata (33). The uv spectrum of the alkaloid (λ_{max} 223 sh, 246, 261 sh, 309, 341 nm; log ε 4.39, 4.66, 4.29, 4.14, 3.66) was suggestive of a 2-methoxycarbazole chromophore (23). The ir spectrum (ν_{max} 1635, 1600, 1480, 1450) lacked an -NH- peak. The nmr spectrum showed signals of H-4 (δ 7.44, s), H-5 (δ 7.86, dd) H-6, H-7, H-8 (δ 7.50–7.0, m) Ar-CH$_3$ (δ 2.30, s), gem-dimethyl (δ 1.26, 1.89, s) O-C-CH$_3$ (δ 1.43, s) and benzylic protons (δ 3.25, m). The physical properties of the alkaloid were in agreement with those of (−) murrayazoline (23) except for its optical rotation. Hence (+) murrayazoline is the antipode of (−) murrayazoline. The absolute configuration of these compounds, like that of (129), is unknown.

(131)

D. Dimeric Carbazoles

1. Bismurrayafoline-A

Bismurrayafoline-A (132), $C_{28}H_{24}N_2O_2$ (M^+ 420), m.p. 176–77°, was obtained as colourless needles from the ethanol extract of *Murraya euchrestifolia* Hayata (*32*). The uv absorption spectrum (λ_{max} 228, 244, 253 sh, 284 sh, 293 nm; log ε 4.81, 4.95, 4.84, 4.15, 4.30) was suggestive of a carbazole skeleton. The nmr spectrum of (**132**) showed signals for an aryl methyl (δ 2.46) and two methoxy groups (δ 3.71 and 3.82). The H-2 and H-2′ protons appeared at (δ 6.62), H-5 at (δ 7.72, d, J = 7 Hz) H-5′ (δ 7.85, d, J = 7 Hz) and the H-4 and H-4′ protons at δ 7.32 and δ 7.36 respectively. Besides it showed signals for eight protons in a complex pattern at (δ 7.0–7.6) and a benzylic methylene (δ 5.83). The mass spectrum showed a molecular ion peak at m/z 420. The base peak was at m/z 210, which was due to half of the molecule and suggested a dimeric carbazole structure for (**132**). Moreover a two proton singlet at δ 5.83 coupled with the observation of the strong peak at m/z 210 suggested the presence of a benzylic methylene group directly bonded to a nitrogen atom.

Bismurrayafoline-A (**132**) on treatment with sodium in liquid ammonia at −75° furnished a compound (**133**) whose nmr spectrum showed the presence of only one methoxy group (δ 3.78) besides other signals. The mass spectrum of (**133**) showed a molecular ion peak at m/z 390. The base peak was at m/z 210. The ohter significant peak was at m/z 181.

(**133**) on methylation afforded an N-methyl derivative (**134**) m.p. 167–169°, whose nmr spectrum confirmed the presence of an N-methyl group (δ 4.04). In the mass spectrum of (**134**) the base peak appeared at m/z 224 and the other peak at m/z 181, as compared to m/z 210 and 181 in the case of (**133**). On this evidence it was concluded that reaction of (**132**) with sodium in liquid ammonia had resulted in substitution by hydrogen of the methoxy group attached to the upper carbazole nucleus instead of in cleavage of the benzylic C-N bond.

Hydrogenolysis of (**132**) with 5% Pd-C in methanol with a small amount of formic acid afforded a compound, identical with murrayafoline-A (**98**) in all respects.

(**132**)

(**133**) R = R₁ = H
(**134**) R = H; R₁ = CH₃

These facts suggested that bismurrayafoline-A is a dimer of (98) with the two halves linked through a carbazole nitrogen of one molecule and the benzylic methylene of the second, as in (132).

2. Bismurrayafoline-B

Bismurrayafoline-B (135), $C_{38}H_{40}N_2O_4$ (M^+ 588), m.p. 260° (dec), was isolated from the acetone extract of *Murraya euchrestifolia* Hayata (32). The ir spectrum (ν_{max} 3550, 3450 and 1615 cm^{-1}) and uv spectrum (λ_{max} 225 sh, 240, 265 sh, 285 sh, 312, 333 sh nm) were suggestive of the presence of carbazole chromophore. On addition of alkali the uv absorption spectrum exhibited a bathochromic shift characteristic of phenols. The mass spectrum of (135) showed the molecular ion peak at m/z 588 which was also the base peak. The appearance of only 19 carbon signals in the ^{13}C nmr spectrum together with the observation of the molecular ion peak at m/z 588 suggested a dimeric structure for bismurrayafoline-B, in which two equivalent monomer units were disposed in a symmetrical manner. The nmr spectrum of bismurrayafoline-B was very similar to that of murrayafoline-B (117) except for the lack of the singlet due to H-2. From these observations it was suggested that bismurrayafoline-B (135) was a dimer of murrayafoline-B (117). Finally the linkage of two murrayafoline-B units at C-2 and C-2′ was confirmed on the basis of data from an NOE experiment leading to structure (135) for bismurrayafoline-B.

(135)

3. (±) Murrafoline

(±) Murrafoline (136), $C_{41}H_{42}N_2O_2$ (M^+ 594), m.p. 260–62°, $[\alpha]_D^{CHCl_3} \pm 0°$, was isolated from the root bark of *Murraya euchrestifolia* Hayata (43). The ir spectrum of the compound (ν_{max} 3450, 1625, 1610 cm^{-1}) and the uv spectrum (λ_{max} 218, 243, 260 sh, 307, 332 sh nm; log ε 4.66, 4.86, 4.66, 4.45, 3.91) was suggestive of a carbazole skeleton.

The nmr spectrum showed signals for two aryl methyls (δ 2.28, 2.37), one vinyl methyl (δ 1.54) and three oxygen-linked tertiary methyl groups (δ 1.40, 1.44) besides a triplet (δ 4.58), a multiplet (δ 3.24) and unresolved overlapping signals for thirteen protons in the region δ 6.70–7.90. The mass spectrum of (136) showed the molecular ion peak at m/z 594. The other significant peak was at m/z 297, which was actually due to half of the molecule and suggested a dimeric structure for (136).

The relative stereochemistry and complete structure of (136) was established by an x-ray analysis.

(136)

E. Carbazolequinones

1. Murrayaquinone-A

Murrayaquinone-A (137), $C_{13}H_9NO_2$ (M^+ 211), m.p. 246–47°, was isolated as brown prism from *Murraya euchrestifolia* (33). The uv spectrum of the compound (λ_{max} 225, 258, 293 sh, 398 nm; log ε 4.63, 4.51, 3.85, 3.93) and ir spectrum (ν_{max} 3200, 1650, 1595 cm^{-1}) indicated the presence of a carbazole-1,4-quinone nucleus in (137) which was confirmed by the appearance of two carbonyl carbon signals at δ 183.4 and 180.4 in the ^{13}C-nmr spectrum. The ^{1}H-nmr spectrum showed the presence of –NH function (δ 9.20), H-6, H-7 and H-8 protons (δ 7.30–7.60). The H-5 proton appeared at δ 8.23 due to the deshielding effect of the carbonyl moiety at C-4. Other signals were those of an

(137)   (138)

allyl methyl group at δ 2.19 and an olefinic proton at δ 6.51. The methyl group was placed at C-3 on the basis of the nmr data and biogenetic considerations.

Photooxidation as well as oxidation of (138) with Fremy's salt furnished a compound which was identical with murrayaquinone-A (137), hence murrayaquinone-A was (137).

2. Murrayaquinone-B

Murrayaquinone-B (118), $C_{19}H_{19}NO_3$ (M^+ 309), m.p. 221–23°, was also isolated from *Murraya euchrestifolia* Hayata (73, 33). The ir spectrum (ν_{max} 3280, 1655, 1640, 1610 cm^{-1}) and uv spectrum (λ_{max} 210 sh, 231, 264, 310 sh, 404 nm; log ε 4.28, 4.58, 4.44, 3.21, 3.66) were suggestive of the presence of a carbazole – 1,4-quinone skeleton. This was confirmed by the appearance of two carbonyl carbon signals at δ 179.8 and δ 183.7. The nmr signals of murrayaquinone-B (118) showed the presence of an -NH proton (δ 9.07), H-5 (δ 7.98, d, J = 9 Hz), H-6 (δ 7.02, d, J = 9 Hz), OCH$_3$ (δ 3.91, s), H-2 (δ 6.42, q, J = 1.5 Hz), an allyl methyl group (δ 2.13), as well as a vinylic proton (δ 5.23), a benzylic methylene (δ 3.6 for two protons) and two methyl groups (δ 1.74 and 1.85) which are parts of an isopentenyl chain. The presence of a methoxy group and the isopentenyl chain was confirmed by the ^{13}C nmr and mass spectra.

The methyl group was placed at C-3 on biogenetic considerations and from the chemical shift of a signal at δ 131.5 in the ^{13}C-nmr spectrum. The location of the methoxyl and prenyl groups at C-7 and C-8, respectively, were settled using an NOE experiment. From all these evidences murrayaquinone-B was formulated as (118).

MARTIN and MOODY (41) have synthesised murrayaquinone-B as follows. 4-(1,1-Dimethylallyloxy)benzaldehyde (139) on condensation with methyl azidoacetate (140) furnished the azidocinnamate (141). This on heating in toluene formed the 6-hydroxyindole (142), which was then converted into the methoxy derivative (143) m.p. 142–43° by treatment with methyl iodide. Claisen condensation of (143) with 4-methylbutyrolactone gave lactone (144). The latter on heating in aqueous dioxane containing a trace of sodium hydroxide formed alcohol (145) m.p. 79–80°, whose reaction with pyridinum chlorochromate formed the corresponding aldehyde (146). The aldehyde on cyclisation with boron trifluoride-methanol complex for 17 hours at room temperature furnished 1,7-dimethoxy-3-methyl-8(3-methylbut-2-enyl)-9H-carbazole (147) whose photooxidation furnished murrayaquinone-B (118) in low yield.

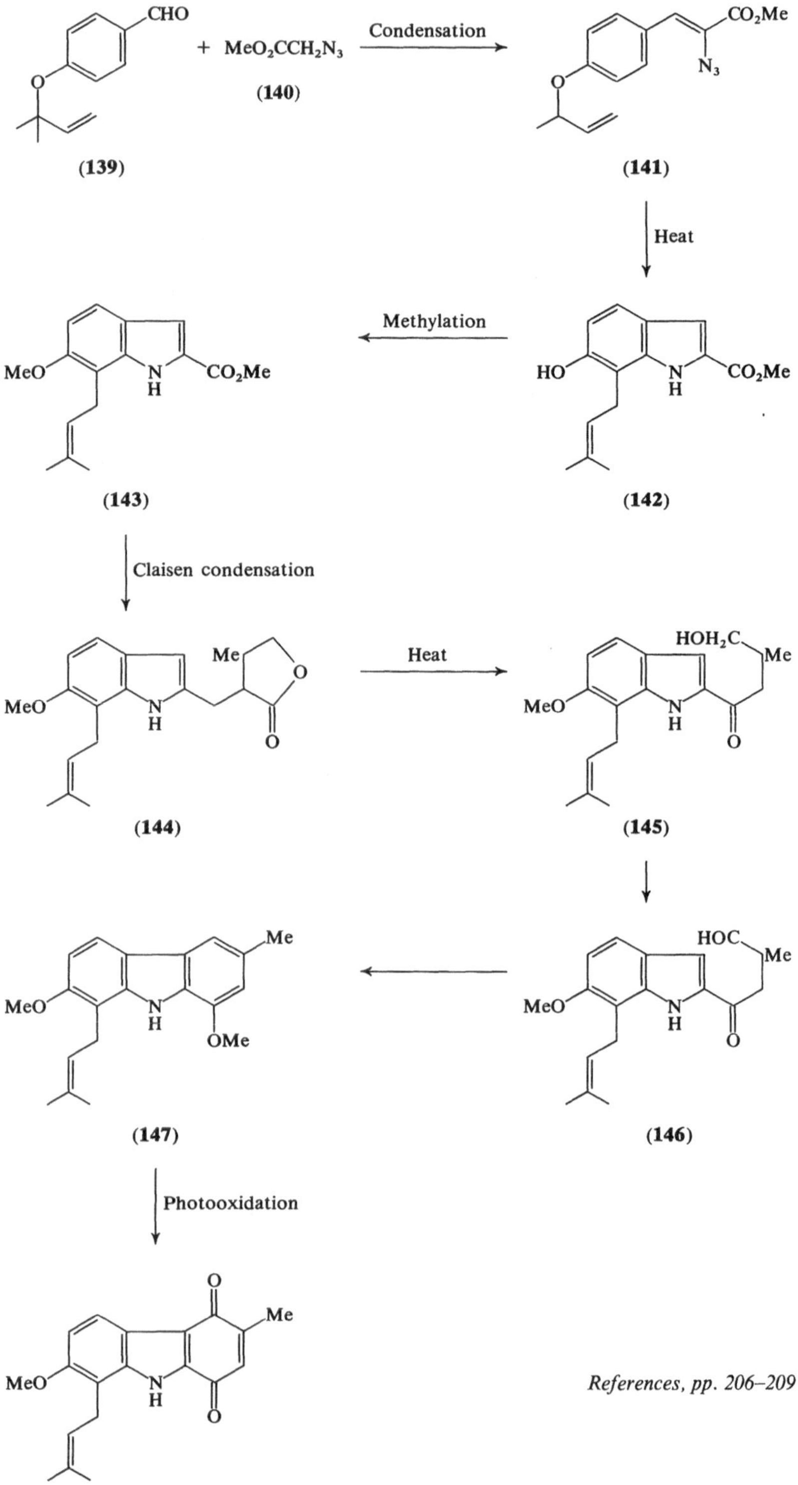

References, pp. 206–209

3. Murrayaquinone-C and Murrayaquinone-D

Murrayaquinone-C (148), $C_{24}H_{27}NO_3$ (M^+ 377), m.p. 158–59°, and murrayaquinone-D (149), $C_{23}H_{25}NO_3$. (M^+ 363), m.p. 164–68°, were isolated from *Murraya euchrestifolia* Hayata (33). The uv absorption spectrum of (148) (λ_{max} 233, 267, 410 nm) and (149) (λ_{max} 215 sh, 234, 266, 415 nm) indicated the presence of same carbazole-1,4-quinone structure in both the compounds. Appearance of two carbonyl carbon signals at δ 183.6 and 179.7 in the ^{13}C nmr spectrum supported this assignment. The nmr spectrum of (148) showed signals for NH (δ 9.08), OCH_3 (δ 3.91), H-2 (δ 6.41), H-5 (δ 7.98), and H-6 (δ 7.01). In addition the presence of a geranyl group [δ 1.56 (3H, s), 1.61 (3H, s), 1.85 (3H, s), 2.05 (4H, s) 3.58 (2H, d, J = 7 Hz), 5.03 (1H, m), and 5.26 (1H, t, J = 7 Hz)] was indicated. The mass spectral fragmentation at m/z 308 [$M^+ - CH_2CH = C(CH_3)_2$] and m/z 254 [$M^+ - CH_2CH = C(CH_3) - CH_2CH = C(CH_3)_2$] also supported the presence of a geranyl group in (148).

Murrayaquinone-D (149) had an nmr spectrum similar to that of (148), the only difference being the presence of an OH signal at (δ 5.52) instead of an OCH_3 signal as in (148). (149) on treatment with diazomethane furnished (148).

The positions of the methoxy and geranyl groups in (148) were settled by an NOE experiment, the structures of murrayaquinone-C and murrayaquinone-D being formulated as (148) and (149), respectively.

(148) R_1=OCH_3 R_2=Geranyl
(149) R_1=OH R_2=Geranyl

F. Carbazoles from Other Sources

1. Hyellazole

Hyellazole (150), $C_{20}H_{17}NO$ (M^+ 287), m.p. 133–34°, was isolated from the blue-green alga *Hyella caespitosa* (22). Its ir (ν_{max} 3490 cm^{-1}) and uv spectrum (λ_{max} 226, 232, 238, 250, 260, 292, 304, 338 and 352 nm; log ε 4.53, 4.55, 4.53, 4.30, 4.13, 4.09, 4.26, 3.65, 3.69) were

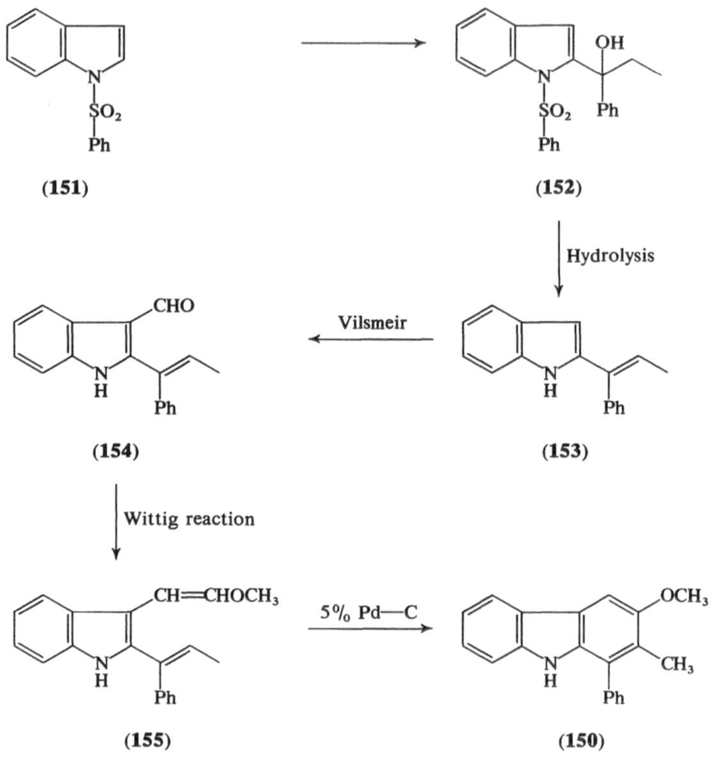

(151)   (152)

Hydrolysis

(154)   Vilsmeir   (153)

Wittig reaction

(155)   5% Pd—C   (150)

suggestive of a carbazole skeleton. The nmr spectrum showed signals at δ 9.52 (brs, 1H, NH), 8.08 (brd, 1H, J=8 Hz; H-5), 7.70, (s, 1H, H-4), 7.6–7.35 (m, 6H, H-8 and five aromatic protons), 7.28 (dd, 1H, J=7.5 and 1 Hz; H-7), 7.11 (dd, 1H, J=7 and 1 Hz; H-6), 3.99 (s, 3H, OCH$_3$) and 2.14 (s, 3H, C-CH$_3$). Using the physical data structure (150) was proposed for hyellazole which was also supported by the ^{13}C nmr spectrum. Structure (150) was confirmed by synthesis (37) as follows. Condensation of N-(benzenesulphonyl)-indole (151) with propiophenone furnished alcohol (152), hydrolysis of which afforded 2-(1-phenyl-1-propenyl)-indole (153). Vilsmeier reaction of (153) gave aldehyde (154). Wittig reaction of (154) gave 2,3-divinylindole (155) which on heating with 5% Pd/C in decalin at 210° furnished hyellazole (150).

2. 6-Chlorohyellazole

6-Chlorohyellazole (156), C$_{20}$H$_{16}$ClNO (M$^+$ 321), m.p. 163–64°, was also isolated from *Hyella caespitosa* (22). The ir and uv data indi-

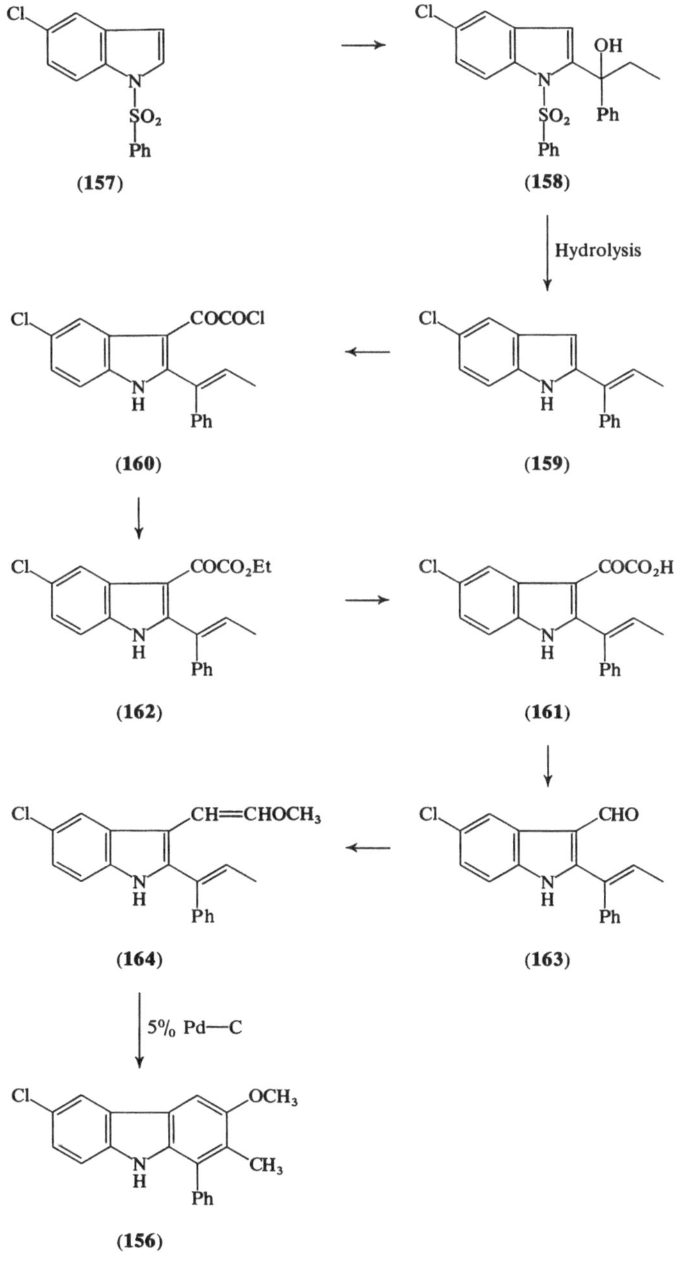

(157)

(158)

Hydrolysis

(160)

(159)

(162)

(161)

(164)

(163)

5% Pd—C

(156)

cated it to be a carbazole alkaloid. The nmr spectrum of (156) indicated that the chlorine was in the 6-position of the carbazole nucleus. The remaining signals in (156) and (150) were identical indicating that the substitution pattern on the other benzenoid ring of the two compounds was identical. The proposed structure (156) was proved by x-ray analysis as well as by synthesis (37). Condensation of N-(benzenesulphonyl)-5-chloroindole (157) with propiophenone furnished alcohol (158) whose hydrolysis yielded 5-chloro-2-(1-phenyl-1-propenyl)-indole (159). (159) with oxalyl chloride gave acid chloride (160), which was converted to keto acid (161) via the ketoester (162). Decarboxylation of (161) gave indole-3-aldehyde (163). Methoxymethylenylation of (163) furnished (164) which on heating with decalin in presence of 5% Pd-C furnished 6-chlorohyellazole (156).

3. Carbazomycin B

Carbazomycin B (54), $C_{15}H_{15}NO_2$ (M^+ 241), m.p. 158.5–160°, was isolated from the cultured mycelium of an unidentified *Streptomyces* sp. designated as strain H 1051-MY 10 (66).

Carbazomycin B gave a positive ferric chloride test. Its uv spectrum (λ_{max} 224, 245, 289, 330 and 340 nm; log ε 4.57, 4.67, 4.20, 3.72 and 3.76) indicated it to be a carbazole derivative. On zinc dust distillation (54) furnished carbazole (2) confirming the presence of a carbazole skeleton. The nmr spectrum showed signals at δ 8.31 (dd, 1H, J = 7.0 and 2.0 Hz; H-5), δ 7.71 (brs, 1H, NH), δ 7.14–7.38 (m, 3H, aromatic protons, H-6, H-7 and H-8), δ 3.80 (s, 3H, OCH₃) in addition to the singlet of a phenolic OH at δ 6.21 and two singlets of two aromatic methyl groups at δ 2.36 and 2.28. Absence of an aromatic proton singlet and the presence of the double doublet at δ 8.31 suggested a 1,2,3,4 tetraasubstituted carbazole structure. This was supported by the ir spectrum which exhibited strong band at v_{max} 750 cm^{-1} due to four adjacent aromatic protons. To locate the position of the hydroxyl group in carbazomycin B, deoxycarbazomycin (165) was prepared by reduction of O-tosylcarbazomycin. The uv spectrum of (165) was closely related to that of 3-methoxy- and 2-methyl-3-methoxycarbazole. In the nmr spectrum of (165) a new aromatic proton singlet appeared at δ 7.45 which was assigned to H-4. From this the hydroxyl was placed at the 4-position.

Carbazomycin B, on acetylation with pyridine and acetic anhydride furnished an acetate (166) and on acetylation by heating with acetic anhydride in the presence of zinc chloride furnished a diacetate (167).

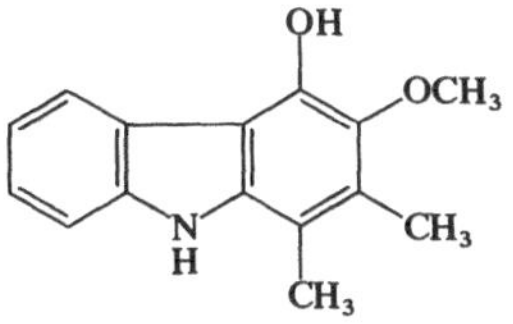

(54)

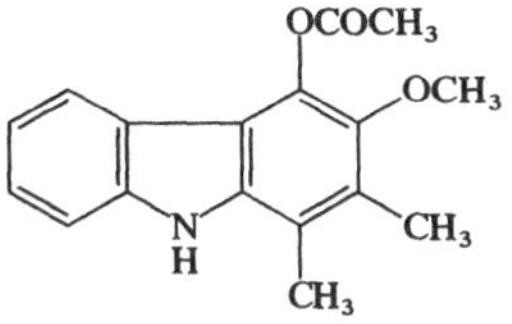

(166)

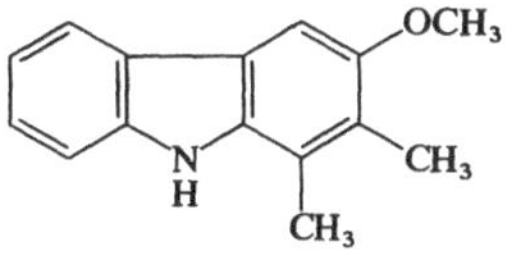

(165)

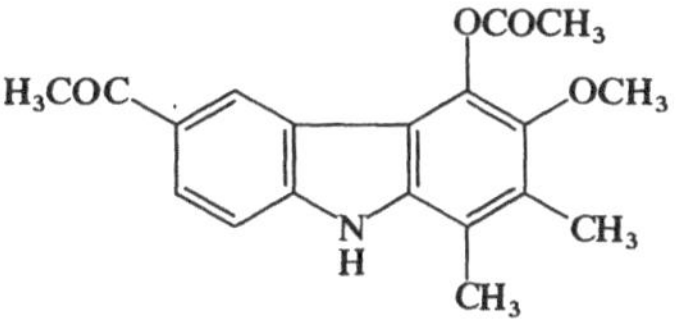

(167)

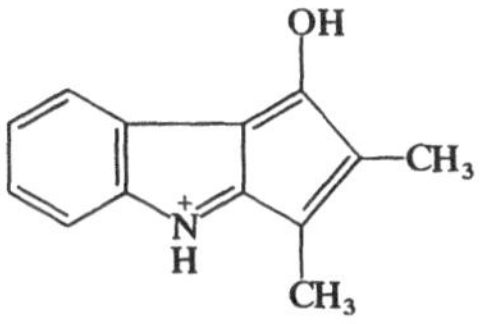

(168)

In the mass spectrum of (54) the characteristic fragmentation peak at m/z 198 (168) proved the presence of the methoxyl group on C-3.

The proposed structure (54) was supported by the ^{13}C-nmr spectrum and confirmed by X-ray crystallographic analysis.

4. Carbazomycin A

Carbazomycin A (53), $C_{16}H_{17}NO_2$ (M$^+$ 255), m.p. 51–52°, was also isolated from the unidentified *Streptomyces* sp. (66). Its uv spectrum (λ_{max} 223, 242, 293, 327 and 340 nm; log ε 4.55, 4.71, 4.31, 3.67

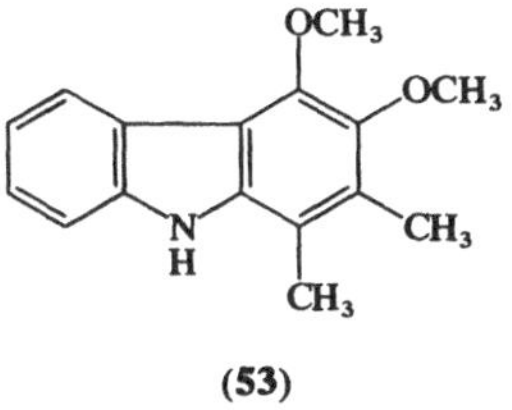

(53)

and 3.66) indicated it to be a carbazole derivative. The nmr spectrum of carbazomycin A was similar to that of carbazomycin B; however the phenolic hydroxyl of (54) at δ 6.21 was replaced by a second methoxyl signal at δ 4.13. Carbazomycin B (54) on methylation with diazomethane furnished carbazomycin A (53). Hence carbazomycin A has structure (53).

References

1. AHOND, A., C. POUPAT, and P. POITIER: Etude Par RM¹³C d'Alacaloides A' Squelette Acridinone 9 (10H) Et Pyrido (4,3b) Carbazole (6H). Tetrahedron **34**, 2385 (1978).
2. AKERMARK, B., L. EBERSON, E. JONSSON, and E. PETTERSSON: Palladium-Promoted Cyclisation of Diphenylether, Diphenylamine and Related Compounds. J. Organ. Chem. (U.S.A.), **40**, 1365 (1975).
3. ALBRECHT, W.L., and R.W. FLEMING: Bis-basic Ethers of Carbazoles: Chem. Abstr. **84**, 150498 s (1976).
4. ALEXANDER, E.J., and A. MOORADIAN: 9-Benzoyl-1,2,3,4-tetrahydro-3-hydroxymethylcarbazole. Chem. Abst. **87**, 39275 q (1977).
5. ASSELIN, A.A., L.G. HUMBER, T.A. DOBSON, J. KOMLOSSY and R.R. MARTEL: Cycloalkanoindoles. 1. Syntheses and Antiinflammatory Actions of some Acidic Tetrahydrocarbazoles, Cyclopentindoles and Cycloheptindoles. J. Medizin. Chem. **19**, 787 (1976).
6. BERGMAN, J., and B. PELCMAN: Synthesis of Carbazoles via 2-Vinylindoles. Tetrahedron Letters 6389 (1985).
7. BERNIER, J.L., J.P. HENICHART, C. VACCHER, and R. HOUSSIN: Condensation of *p*-Benzoquinone with 4-Cyano and 4-Nitroanilines. An Extension of the Nenitzescu Reaction. J. Organ. Chem. (U.S.A.) **45**, 1493 (1980).
8. BHATTACHARYYA, L., S.K. ROY, and D.P. CHAKRABORTY: Structure of the Carbazole Alkaloid Isomurrayazoline from *Murraya koenigii*. Phytochem. **21**, 2432 (1982).
9. BHATTACHARYYA, P., and A. CHAKRABORTY: Mukonal, a Probable Biogentic Intermediate of Pyranocarbazole Alkaloids from *Murraya koenigii*. Phytochem. **23**, 471 (1984).
10. BHATTACHARYYA, P., A. CHAKRABORTY, and B.K. CHOWDHURY: A New Synthesis of Girinimbine. Indian. J. Chem. **23 B**, 849 (1984).
11. – – – Heptazolicine, a Carbazole Alkaloid from *Clausena heptaphylla*. Phytochem. **23**, 2409 (1984).
12. BHATTACHARYYA, P., and B.K. CHOWDHURY: 2-Methoxy-3-methyl Carbazole from *Murraya koenigii*. Indian J. Chem. **24 B**, 452 (1985).
13. – – Glycozolidal, a new Carbazole Alkaloid from *Glycosmis pentaphylla*. J. Natural Products **48**, 465 (1985).
14. – – Clausenapin: a new Carbazole Alkaloid from *Clausena heptaphylla* Wt. & Arn. Chem. and Ind. 301 (1984).
15. BHATTACHARYYA, P., P.K. CHAKRABARTTY, and B.K. CHOWDHURY: Glycozolidol, an Antibacterial Carbazole Alkaloid from *Glycosmis pentaphylla*. Phytochem. **24**, 882 (1985).
16. BHATTACHARYYA, P., and S.S. JASH: Benzoylperoxide as a Spray reagent for Carbazole Alkaloids. J. Chromatogr. **298**, 200 (1984).
17. BHATTACHARYYA, P., S.S. JASH, and A.K. DEY: Free radical Cyclisation of Diphenylamine: a Convenient Synthesis of Carbazole and 3-Methylcarbazole. J.C.S. Chem. Commun. 1668 (1984).
18. BHATTACHARYYA, P., S.S. JASH, and B.K. CHOWDHURY: A Biogenetically important Carbazole Alkaloid from *Murraya koenigii* Spreng. Chem. and Ind. 246 (1986).
19. BHATTACHARYYA, P., T. SARKAR, A. CHAKRABORTY, and B.K. CHOWDHURY: Structure

and Synthesis of Glycozolinol, a new Carbazole Alkaloid from *Glycosmis pentaphylla* (Retz) D.C. Indian J. Chem. **23 B**, 49 (1984).

20. BORDNER, J., D.P. CHAKRABORTY, B.K. CHOWDHURY, S.N. GANGULY, K.C. DAS, and B. WEINSTEIN: The x-ray Crystal Structure of Murrayazoline (Mahanimbidine and Currayangin). Experientia **28**, 1406 (1972).

21. BOWEN, I.H., and K.P.W. CHRISTOPHER PERERA: Alkaloids, Coumarins and Flavonoids of *Micromelum zeylanicum*. Phytochem. **21**, 433 (1982).

22. CARDELLINA, J.H. II., M.C. KIRKUP, R.E. MOORE, J.S. MYNDERSE, K. SEFF, and C.J. SIMMONS: Hyellazole and Chlorohyellazole, two Novel Carbazoles from the Bluegreen Alga *Hyella Caespitosa* Born. et. Flah. Tetrahedron Letters, 4915 (1979).

23. CHAKRABORTY, D.P.: Carbazole Alkaloids. In: Fortschritte der Chemie Organischer Naturstoffe, Vol. 34 (W. HERZ, H. GRIESBACH, and G.W. KIRBY, Eds), pp. 299–371. Wien-New York: Springer 1977.

24. CHAKRABORTY, D.P., A. ISLAM, and S. ROY: 2-Methylanthraquinone from *Clausena heptaphylla*. Phytochem. **17**, 2043 (1978).

25. CHAKRABORTY, D.P., B.K. BARMAN, and P.K. BOSE: On the Constitution of Murrayanine, a Carbazole Derivative Isolated from *Murraya koenigii* Spreng. Tetrahedron **21**, 681 (1965).

26. CHAKRABORTY, D.P., S. ROY, and R. GUHA: Structure of Mukonidine. J. Indian Chem. Soc. **55**, 1114 (1978).

27. CHOWDHURY, B.K., A. MUSTAPHA, and P. BHATTACHARYYA: Separation of Carbazole Alkaloids by Gas-liquid Chromatography. J. Chromatogr. **329**, 178 (1985).

28. CHOWDHURY, B.K., S.K. HIRANI, and P. BHATTACHARYYA: Highperformance Liquid Chromatographic Separation of Carbazole Alkaloids. J. Chromatogr. **369**, 258 (1986).

29. CHOWDHURY, D.N., S.K. BASAK, and B.P. DAS: Studies on the Insecticidal and Antimicrobial Properties of Some Carbazole Alkaloids. Current Sci. (India) **47**, 490 (1978).

30. DIFFERDING, E., and L. GHOSEZ: Intramolecular Diels-Alder Cycloaddition of Vinylketenimines. A Convergent Route to Carbazoles and Pyridocarbazole Alkaloids. Tetrahedron Letters 1647 (1985).

31. FIEBIG, M., JOHN M, PEZZUTO, DJAJA D, SOEJARTO and A. DOUGLAS KINGHORN: Koenoline, a further Cytotoxic Carbazole Alkaloid from *Murraya koenigii*. Phytochem. **24**, 3041 (1985).

32. FURUKAWA, H., T.S. WU, and T. OHTA: Bismurrayafoline-A and B, two Novel Dimeric Carbazole Alkaloids from *Murraya euchrestifolia*. Chem. Pharm. Bull. **31**, 4202 (1983).

33. FURUKAWA, H., T.S. WU, T. OHTA, and C.S. KUOH: Chemical Constituents of *Murraya euchrestifolia* Hayata. Structures of Novel Carbazolequinones and other New Carbazole Alkaloids. Chem. Pharm. Bull. **33**, 4132 (1985).

34. GANGULY, S.N., and A. SARKAR: Exozoline, a New Carbazole Alkaloid from the Leaves of *Murraya exotica*. Phytochem. **17**, 1816 (1978).

35. JOSHI, B.S., V.N. KAMAT, A.K. SAKSENA, and T.R. GOVINDACHARI: Structure of Heptaphylline, a Carbazole Alkaloid from *Clausena heptaphylla* Wt. & Arn. Tetrahedron Letters 4019, 1967.

36. JOSHI, B.S., V.N. KAMAT, and D.H. GAWAD: On the Structure of Girinimbine, Mahanimbine, Isomahanimbine, Koenimbidine and Murrayacine. Tetrahedron **26**, 1475 (1970).

37. KANO, S., E. SUGINO, S. SHIBUYA, and S. HIBINO: Synthesis of Carbazole Alkaloids Hyellazole and 6-Chlorohyellazole. J. Organ. Chem. (U.S.A.) **46**, 3856 (1981).

38. KAPIL, R.S.,: Carbazole Alkaloids, in Alkaloids, Vol. 13 (R.H.F. MANSKE, Ed.), pp. 273. New York – London: Academic Press, Inc. 1971.

39. KONG, Y.C., K.K. NG, H.P.P. BUT, L. QUIAN, Y. SIXAO, H.T. ZHANG, K.F. CHENG, D.D. SOEJARTO, K.W. SONG and P.G. WATERMAN: Sources of the Anti-implantation Alkaloid Yuechukene in the Genus *Murraya*. J. Ethnopharmacol. **15**, 195 (1986).

40. LONTSI, D., J.F. AYAFOR, B.L. SONDENGAM, J.D. CONOLLY, and D.S. RYCROFT: The use of two Dimensional Long-range δ_C/δ_H correlation in Conjunction with the One-Dimensional Protoncoupled ^{13}C NMR Spectrum in the Structural Elucidation of Ekeberginine, a New Carbazole Alkaloid from *Ekebergia senegalensis* (Meliaceae). Tetrahedron Letters 26, 4249 (1985).

41. MARTIN, T., and C.J. MOODY: Synthesis of the Carbazole Alkaloid Murrayaquinone – B. J.C.S. Chem. Commun. 1391 (1985).

42. MASAKI, M., T. YUSHIRO, S. TSUTOMU, and N. KEIJU: Pharmacological Studies on Carprofen, a new Non-steroidal Anti-inflammatory Drug in Animals. Nippon Kagaku Zasshi. 73, 757 (1977).

43. MCPHAIL, A.T., T.S. WU, T. OHTA, and H. FURUKAWA: Structure of ($\pm$)-Murrafoline, a Novel Biscarbazole Alkaloid from *Murraya euchrestifolia*. Tetrahedron Letters 5377 (1983).

44. MESTER, I., D. BERGENTHAL, and J. REISCH: Inhaltsstoffe aus *Clausena anisata* (Willd) Oliv. (Rutaceae). III. ^{13}C-NMR Spektren des Mupamins, des Carbazols und einiger Carbazol Derivative. Z. Naturforsch. 34 b, 650 (1979).

45. MESTER, I., and J. REISCH: Isolierung und Struktur des Mupamins, eines Neuen Carbazol Alkaloids. Liebigs Ann. Chem. 1925 (1977).

46. MESTER, I., M.K. CHOUDHURY, and J. REISCH: Synthese des Mupamins. Liebigs Ann. Chem. 241 (1980).

47. MOORADIAN, A., P.E. DUPONT, A.G. ALAVAC, M.D., and J. PEARL: 3-Aminotetrahydrocarbazoles as a New Series of Central Nervous System Agents. J. Medizin. Chem. 20, 487 (1977).

48. MUKHERJEE, M., S. MUKHERJEE, A.K. SHAW, and S.N. GANGULY: Mukonicine, a Carbazole Alkaloid from leaves of *Murraya koenigii*. Phytochem. 22, 2328 (1983).

49. MUKHERJEE, S., M. MUKHERJEE, and S.N. GANGULI: Glycozolinine, a Carbazole Derivative from *Glycosmis pentaphylla*. Phytochem. 22, 1064 (1983).

50. NAKAMURA, S.: Paper Presented at the National Symposium on Natural Product Chemistry, Feb. 1983 at Bose Institute (Calcutta).

51. NARASIMHAN, N.S., M.V. PARADHKAR, and S.L. KELKAR: Alkaloids of *Murraya koenigii*. Structure of Mahanine, Koenine, Koenigine, and Koenidine. Indian J. Chem. 8, 473 (1970).

52. NARASIMHAN, N.S., M.V. PARADHKAR, and V.P. CHITGUPPI: Structures of Mahanimbine and Koenimbine. Tetrahedron Letters 5502 (1968).

53. NOLAND, E.E., and S.R. WANN: In Situ Vinyl Indole Synthesis of Carbazoles. J. Organ. Chem. (U.S.A.). 44, 4402 (1979).

54. OIKAWA, Y., and O. YONEMITSU: A New Synthetic Method for Condensed Heterocycles, Carbazoles, Indoles and Benzothiophenes Based on Acid-Catalysed Cyclisation of β-Ketosulfoxides. J. Organ. Chem. (U.S.A.). 41, 1118 (1976).

55. OIKAWA, Y., and Y. OSAMU: Reactions and Synthetic Applications of β-Ketosulfoxides VII. A Novel Synthesis of Pyranocarbazole Alkaloids Girinimbine and Murrayacine. Heterocycles 233 (1976).

56. OKORIE, D.A.: A New Carbazole Alkaloid and Coumarins from Roots of *Clausena anisata*. Phytochem. 14, 2720 (1975).

57. PECCA, J.G., and S.M. ALBONICO: Synthetic Trypanocides I. Substituted 1,2,3,4-Tetrahydrocarbazoles. J. Medizin. Chem. 13, 327 (1970).

58. PRAKASH, D., K. RAJ, R.S. KAPIL, and S.P. POPLI: Chemical Constituents of *Clausena lansium*: Part I – Structure of Lansamide I and Lansine. Indian J. Chem. 13 B, 1075 (1980).

59. RAMARAO, A.V., K.S. BHIDE, and R.B. MAZUMDAR: Mahanimbinol. Chem. and Ind. 697 (1980).

60. RANDELLA, B.E., and B.P.J. PATEL: Antimicrobial Activity of Carbazole Derivatives. Experientia 38, 529 (1982).

61. RICE, L.M., and K.R. SCOT: 3-Substituted 1,2,3,4-tetrahydrocarbazoles. J. Medizin. Chem. **13**, 308 (1970).
62. ROY, S., L. BHATTACHARYYA, and D.P. CHAKRABORTY: Structure and Synthesis of Mukoline and Mukolidine, Two new Carbazole Alkaloids from *Murraya koenigii* Spreng. J. Indian Chem. Soc. **59**, 1369 (1982).
63. ROY, S., R. GUHA, S. GHOSH, and D.P. CHAKRABORTY: Biomimetic Hydroxylation Studies on Carbazole Alkaloids. Indian J. Chem. **21 B**, 617 (1982).
64. ROY, S., S. GHOSH, and D.P. CHAKRABORTY: Thin-layer Chromatographic Studies with Some Carbazoles: Hydrochloric acid – a Convenient Spray Reagent for Carbazole Alkaloids. J. Indian Chem. Soc. **58**, 296 (1981).
65. – – – Structure of Mahanimboline. Chem. and Ind. 669 (1979).
66. SAKANO, K., and S. NAKAMURA: New Antibiotics Carbazomycins A and B II. Structural Elucidation. J. Antibiotics **33**, 961 (1980).
67. SAROJA, B., and P.C. SRINIVASAN: A Simple Route to Indole-2,3-quinodimethane – a Facile Synthesis of Carbazoles. Tetrahedron Letters 5429 (1984).
68. SHARMA, R.B., and R.S. KAPIL: Synthesis of Lansine and 6-Methoxyheptaphylline. Chem. and Ind. 158 (1980).
69. – – Synthesis of Heptazoline. Chem. and Ind. 268 (1982).
70. SHARMA, R.B., (Mrs.) F. ANWER, and R.S. KAPIL: Synthesis of Mupamine. Indian J. Chem. **20 B**, 701 (1981).
71. SHOEB, A., F. ANWER, R.S. KAPIL, S.P. POPLI, P.R. DUA, and B.N. DHAWAN: N-Alkylamino Carbazoles as Potent Anticonvulsant and Diuretic Agents. J. Medizin. Chem. **16**, 425 (1973).
72. SNOOK, M.E., R.F. ARRENDALE, H.C. HIGMAN, and O.T. CHORTYK: Isolation of Indoles and Carbazoles from Cigarette Smoke Condensate. Anal. Chem. **50**, 88 (1978).
73. WU, T.S., T. OHTA, and H. FURUKAWA: Structure of Murrayaquinone-B, a Novel Carbazole Alkaloid from *Murraya euchrestifolia* Hayata. Heterocycles **20**, 1267 (1983).

(*Received January 9, 1987*)

Author Index

Page numbers printed in *italics* refer to References

Subject Index

Composition: Universitätsdruckerei H. Stürtz AG, D-8700 Würzburg
Printed by novographic, Ing. W. Schmid, A-1238 Wien

Fortschritte der Chemie organischer Naturstoffe
Progress in the Chemistry of Organic Natural Products

Volume 51:

1987. VII, 317 pages. Cloth DM 280,–, öS 1960,–.
ISBN 3-211-81972-X
Contents: M. Gill and W. Steglich: Pigments of Fungi (Macromycetes).

Volume 50:

1986. 71 figures. IX, 261 pages. Cloth DM 210,–, öS 1470,–.
ISBN 3-211-81969-X
Contents: L. Jaenicke and F.-J. Marner: The Irones and Their Precursors. –
M. Lounasmaa and P. Somersalo: The Condylocarpine Group of Indole
Alkaloids. – U. Séquin: The Antibiotics of the Pluramycin Group (4*H*-Anthra
[1,2-*b*]pyran Antibiotics). – R. M. Wenger: Cyclosporine and Analogues –
Isolation and Synthesis – Mechanism of Action and Structural Requirements
for Pharmacological Activity. – H. Inouye and S. Uesato: Biosynthesis of
Iridoids and Secoiridoids.

Volume 49:

1986. VIII, 400 pages. Cloth DM 290,–, öS 2030,–. ISBN 3-211-81910-X
Contents: R. A. Hill: Naturally Occurring Isocoumarins. – R. Wijnsma and
R. Verpoorte: Anthraquinones in the Rubiaceae. – H. Chr. Krebs: Recent
Developments in the Field of Marine Natural Products with Emphasis on
Biologically Active Compounds.

Volume 48:

1985. 33 figures. IX, 285 pages. Cloth DM 220,–, öS 1540,–.
ISBN 3-211-81886-3
Contents: P. S. Steyn and R. Vleggaar: Tremorgenic Mycotoxins. –
R. E. Moore: Structure of Palytoxin. – P. Crews and S. Naylor: Sesterterpenes:
An Emerging Group of Metabolites from Marine and Terrestrial Organisms.

Volume 47:

1985. 16 figures. VIII, 290 pages. Cloth DM 198,—, öS 1390,—.
ISBN 3-211-81864-2
Contents: R. Southgate and S. Elson: Naturally Occurring β-Lactams. — I. Howe and M. Jarman: New Techniques for the Mass Spectrometry of Natural Products. — P. G. McDougal and N. R. Schmuff: Chemical Synthesis of the Trichothecenes. — J. Polonsky: Quassinoid Bitter Principles II.

Volume 46:

1984. 7 figures. IX, 253 pages. Cloth DM 178,—, öS 1250,—.
ISBN 3-211-81804-9
Contents: O. Tanaka and R. Kasai: Saponins of Ginseng and Related Plants. — E. Fujita and M. Node: Diterpenoids of *Rabdosia* Species. — S. Johne: The Quinazoline Alkaloids.

Volume 45:

1984. 2 figures. VIII, 288 pages. Cloth DM 194,—, öS 1360,—.
ISBN 3-211-81755-7
Contents: D. A. H. Taylor: The Chemistry of the Limonoids from Meliaceae. — J. A. Elix, A. A. Whitton, and M. V. Sargent: Recent Progress in the Chemistry of Lichen Substances. — Y. Shimizu: Paralytic Shellfish Poisons.

All Volumes and Cumulative Index 1—20 available

Price reduction for subscribers: 10%

Special reduced price (20% reduction) for the complete Series Vols. 1—51 incl. the Cumulative Index to Vols. 1—20

Springer-Verlag Wien New York

Mölkerbastei 5, A-1011 Wien
175 Fifth Avenue, New York, NY 10010, U.S.A.
Heidelberger Platz 3, D-1000 Berlin 33
37-3, Hongo 3-chome, Bunkyo-ku, Tokyo 113, Japan